T0135460

Aram Hajian, Nelson Baloian, Tomoo Inoue, Wolfram Luther (Eds.)

Data Science, Human-Centered Computing, and Intelligent Technologies

Logos Verlag Berlin

Bibliographic information published by Die Deutsche Bibliothek

Die Deutsche Bibliothek lists this publication in the Deutsche Nationalbibliografie; detailed bibliographic data is available in the Internet at http://dnb.d-nb.de.

ISBN 978-3-8325-5520-7

Logos Verlag Berlin GmbH
Georg-Knorr-Str. 4, Geb. 10,
12681 Berlin

Tel.: +49 (0)30 / 42 85 10 90
Fax: +49 (0)30 / 42 85 10 92
http://www.logos-verlag.de

Aram Hajian, Nelson Baloian, Tomoo Inoue, Wolfram Luther (Eds.)

Data Science, Human-Centered Computing, and Intelligent Technologies

3rd International Workshop at the American University of Armenia, College of Science & Engineering

August 23–25, 2022

Co-organized with IEEE Computer Society Armenia Chapter

Revised contributions

Volume Editors

Aram Hajian
American University of Armenia, College of Science and Engineering
40 Marshal Baghramyan Ave, Yerevan 0019, Armenia
Email: ahajian@aua.am

Nelson Baloian
Department of Computer Science, Universidad de Chile
Blanco Encalada 2120, Santiago 6511224, Chile
E-mail: nbaloian@dcc.uchile.cl

Tomoo Inoue
University of Tsukuba, Faculty of Library, Information and Media Science
1-2, Kasuga, Tsukuba, Ibaraki, 305-8550, Japan
E-mail: inoue@slis.tsukuba.ac.jp

Wolfram Luther
University of Duisburg-Essen, Scientific Computing, Computer Graphics, and Image Processing
Lotharstraße 63, 47057 Duisburg, Germany
E-mail: wolfram.luther@uni-due.de

ACM Subject Classification (1998): C.4, E.2-4, H.1, H.5.2-4, I.2, I.4-6, J.1, J.3-5

The Open Access publication of this title under CC BY-NC-SA 4.0 license was made possible with the support from the Publication Fund of the University of Duisburg-Essen.

PREFACE

After two successful versions of the Workshop on Collaborative Technologies and Data Science in Smart City Applications (CODASSCA 2018 and 2020) we are glad to present proceedings of the third version held at the American University of Armenia (AUA), August 23–25, 2022 in Yerevan, Armenia.

This book presents 15 selected and carefully revised papers, which were originally made available at the start of the workshop in the Open Access Proceedings, also entitled Data Science, Human-Centered Computing, and Intelligent Technologies.

Society, technology, and science are undergoing a rapid and revolutionary transformation towards incorporating Artificial Intelligence in every system humans use in everyday life for creating Smart Environments (SmE) through Ambient Intelligence (AmI) in highly interconnected and collaborative scenarios. The main source and asset for making smart systems is rich data, which is produced today in extraordinarily large quantities thanks to recent advances in sensors and sensor networks and is carefully processed for pervasive and embedded computing. Rich data enhances the capabilities of everyday objects and eases collaboration among people.

Mobile systems could enhance the possibilities available for designers and practitioners. Effective analysis, quality assessment, and utilization of big data are key factors for success in many business and service domains, including smart systems.

Major industrial domains are on the way to performing this tectonic shift based on Big Data, Artificial Intelligence, Collaborative Technologies, Smart Environments supporting Virtual and Mixed Reality Applications, Multimodal Interaction, and Reliable Visual and Cognitive Analytics.

However, before we can effectively and efficiently turn the huge amount of generated data into information and knowledge, a number of requirements must be fulfilled and international standards for the quality of and access to the data developed and applied. The first requirement is to ensure that data quality—which includes the accuracy and integrity of the obtained data, timely delivery, suitable quantity, integrity, privacy and security requirements, and Digital Rights Management—complement realization and deployment of modern design, implementation, and evaluation tools. The second is to develop models, which can turn the data into valuable information and then into knowledge. Two important characteristics are desirable for regression and classification models: accuracy and interpretability. While accuracy deals with the ability of the model to predict a certain outcome, interpretability deals with the ability of the model to explain the reasons for producing a certain outcome. The aim of this workshop is to bring together researchers and practitioners working on both theoretical and practical aspects of data generation, data processing, and knowledge creation. These aspects include social issues that arise when using AI-powered systems in collaborative scenarios and smart cities applications. The following two sections each contain seven contributions, which we introduce below.

Data Science and Intelligent Technologies

1 In "Information-Theoretic Investigation of Authenticated Steganographic Model in the Presence of Active Adversary" Mariam Haroutunian, Parandzem Hakobyan, Ashot Harutyunyan, and Arman Avetisyan consider an information-theoretic model of a stegosystem with an active adversary and two principal communication modes: covertext and stegotext. For stegotext, the task is to determine the legitimacy of the sender using two-stage statistical hypothesis testing. For the receiver side, an asymptotically optimal testing approach in a logarithmic scale is used to examine and specify functional dependence of errors in both stages.

2 Sergey Abrahamyan proposes "Another Approach for ZKRP Algorithms" dealing with a zero-knowledge range proof that a secret integer belongs to a certain interval without conveying any additional information. Based on an order-revealing encryption scheme introduced by Wu and Lewi in 2016, modifications in the ZKRP scheme are made which ensure completeness and soundness and minimize dependence on a trusted setup. The performance data, security parameters, and other details will be presented in the full paper.

3 In "Preserving Location and Query Privacy Using a Broadcasting LBS" the authors Pablo Torres, Patricio Galdames, Claudio Gutierrez-Soto, and Geraldine Solar consider a location and query privacy-aware LBS server intended to proactively broadcast location-based data periodically to users moving in a specific geographic area. This research develops ideas to establish a balance between privacy issues and response time for users with changing interests.

4 The authors Ashot Harutyunyan, Nelli Aghajanyan, Lilit Harutyunyan, Arnak Poghosyan, Tigran Bunarjyan, and A.J. Han Vinck present "On Diagnosing Cloud Applications with Explainable AI." In it, they discuss experimental modeling and evaluation of cloud monitoring and management software for complete KPI diagnosis and interpretability. They break down the process into three steps: (1) developing and applying a KPI metric to create multiple quality class IDs for labeling an entire application dataset, (2) training regression and classification models to optimize the KPI behavior and evaluate the components that can be used for explanatory purposes, and (3) including the application of decision trees and rule induction algorithms as a form of explainable AI to derive consistent conditions for KPI failures.

5 In "Root Cause Analysis of Application Performance Degradations via Distributed Tracing," Arnak Poghosyan, Ashot Harutyunyan, Naira Grigoryan and Clement Pang explore the application of rule learning classification algorithms to a distributed tracing of traffic with the aim of identifying potential conditions that explain malfunctioning services via trace types, spans, and dimensions.

6 According to the authors Ashot Baghdasaryan, Tigran Bunarjyan, Arnak Poghosyan, Ashot Harutyunyan, and Jad El-Zein "On AI-Driven Customer Support in Cloud Operations" addresses the development of proactive and intelligent AI-driven analytics that not only automate and accelerate the resolution of currently available system requests, but also anticipate potential malfunctioning through proper customer segmentation and rule discovery.

7 Lilit Yolyan gives an overview on "Computer Vision Application for Smart Cities Using Remote Sensing Data." The application of AI and computer vision techniques are important issues in solving smart city tasks such as surveillance, area coverage, land usage and coverage, damage monitoring, and fire detection. In her paper, Yolyan analyzes those methods using remote sensing data from the city of Yerevan.

Human-Centered Computing and Intelligent Technologies

8 Zhen He, Sayan Sarcar, and Tomoo Inoue present "Exploring the Feasibility of Video Activity Reporting for Students in Distance Learning." In a distributed labor mode, text-based reporting is insufficient to manage and evaluate individuals' work progress and engagement. Video reporting is considered a way to improve the situation. This paper offers the authors' initial observations of an experimental setup on video activity reporting.

9 The authors Ari Nugraha and Tomoo Inoue report on "An Experiment of Crowd-sourced Online Collaborative Question Generation and Improvement for Video Learning materials in Higher Education." They use the Amazon Mechanical Turk platform to generate and collaboratively improve questions through multiple refinement processes related to video lectures as learning material.

10 Davit Karamyan, Ara Abovyan, and Tigran Karamyan consider and develop "Compact N-gram Language Models for Armenian." Trade-offs between model size and perplexity measure are investigated. Authors also release compact language models trained on very large corpora. They examine the impact of pruning and quantization methods on model size reduction and use Byte Pair Encoding. Their result is a compact subword language model trained on huge Armenian corpora.

11 Maximilian Khotilin, Rustam Paringer, Alexander Kupriyanov, Dmitriy Kirsh, David Asatryan, Mariam Haroutunian, and Artem Mukhin highlight the "Estimation of Hyperspectral Images Bands Similarity Using Textural Properties." Textural properties of given image classes using discriminant analysis techniques are the basis for generating a set of effective features that help analyze the properties of hyperspectral images. Afterwards, visual analytics and knowledge creation provide insights in areas such as agriculture, steel plate quality, etc.

12 "Recognition of Convex Bodies by Probabilistic Methods" by Victor Ohanyan is a survey contribution summarizing results obtained by a research group in tomography of planar bounded convex bodies $D \subset \mathbf{R}^n$ during the last 20 years. Since the estimates of probability characteristics can be obtained using well-known methods of mathematical statistics, reconstruction of convex bodies using random sections such as chord (length) or k-dimensional random flats with $n>k\geq 2$ makes it possible to simplify the approach.

13 "Validation of Risk Assessment Models for Breast and Ovarian Cancer–Related Gene Variants" is the title of a contribution by Wolfram Luther. The paper focuses on breast and ovarian cancer–related gene variants risk assessment, related models, and their validation and standardization. Based on user input data, such methods test for the occurrence in individuals and their relatives of deleterious gene variants that have been

shown to be highly responsible for these cancer types. Two ways of mathematically modeling the gene mutation risk are proposed. They are based on individuals' health status and family history of disease. The Dempster-Shafer theory of evidence plays a key role in model evaluations.

14 The authors Nelson Baloian, Wolfram Luther, Sergio Peñafiel, and Gustavo Zurita bring together their experiences and recommendations on "Evaluation of Cancer and Stroke Risk Scoring Online Tools" in a joint contribution about the evaluation of several public services on cancer and stroke risk scoring. Risk scoring is based on users' input data concerning their own and their families' health parameters, behaviors, and diseases. The authors discuss performance and interpretability aspects of those services and derive minimum requirements for such risk scoring tools.

15 A contribution by Nelson Baloian, Belisario Panay, Sergio Peñafiel, José A. Pino, Jonathan Frez, and Cristóbal Fuenzalida entitled "Sales Goals Planning using Evidence Regression" introduces an intelligent machine to help sales goal planners who manage brick-and-mortar retail stores to achieve their objectives through model tuning. With reference to the Dempster-Shafer theory of evidence, the authors elaborate on regression and classification settings.

The editors would like to express their gratitude to the German Research Foundation (DFG) and the German Academic Exchange Service (DAAD) for funding their activities; to Rubina Danilova, Ashot Harutyanyan, and Gregor Schiele for their ongoing encouragement and support; and to all participants for their presentations and contributions to the workshop and this proceedings volume.

Yerevan, Tsukuba, Duisburg, August 2022

The Editors: Aram Hajian, Nelson Baloian, Tomoo Inoue, and Wolfram Luther

Technical Committees

Organizing Committee Chair: Rubina Danilova (Armenia), Aram Hajian (Armenia), Nelson Baloian (Armenia, Chile), Gregor Schiele (Germany)

Program Committee Chair: Ashot Harutyunyan (Armenia), José A. Pino (Chile), Wolfram Luther (Germany), Tomoo Inoue (Japan)

Program Chair Track Intelligent Technologies and Data Science: Ashot Harutyunyan (Armenia)

Program Chair Track Collaborative Technologies: Nelson Baloian (Chile)

Program Chair Track Smart Human-Centered Computing: Tomoo Inoue (Japan)

Contents

3rd Workshop on Collaborative Technologies and Data Science in Smart City Applications
A. Hajian, N. Baloian, T. Inoue, and W. Luther v

Contents x

Data Science and Intelligent Technologies

Information-Theoretic Investigation of Authenticated Steganographic Model in the Presence of Active Adversary
M. Haroutunian, P. Hakobyan, A. Harutyunyan, and A. Avetisyan 1

Another approach for ZKRP algorithms
S. Abrahamyan 8

Preserving Location and Query Privacy Using a Broadcasting LBS
P. Torres, P. Galdames, C. Gutierrez-Soto, and G. Solar 11

On Diagnosing Cloud Applications with Explainable AI
A. Harutyunyan, N. Aghajanyan, L. Harutyunyan, A. Poghosyan, T. Bunarjyan, and A.J. Han Vinck 23

Root Cause Analysis of Application Performance Degradations via Distributed Tracing
A. Poghosyan, A. Harutyunyan, N. Grigoryan, and C. Pang................... 27

On AI-Driven Customer Support in Cloud Operations
A. Baghdasaryan, T. Bunarjyan, A. Poghosyan, A. Harutyunyan, and J. El-Zein 32

Computer Vision Application for Smart Cities Using Remote Sensing Data: Review
L. Yolyan 36

HUMAN-CENTERED COMPUTING AND INTELLIGENT TECHNOLOGIES

Exploring the Feasibility of Video Activity Reporting for Students in Distance Learning
Z. He, S. Sarcar, and T. Inoue ... 44

An Experiment of Crowdsourced Online Collaborative Question Generation and Improvement for Video Learning Materials in Higher Education
A. Nugraha, T. Inoue ... 56

Compact N-gram Language Models for Armenian
D. Karamyan, A. Abovyan, and T. Karamyan ... 67

Estimation of Hyperspectral Images Bands Similarity Using Textural Properties
M. Khotilin, R. Paringer, A. Kupriyanov, D. Kirsh, D. Asatryan, M. Haroutunian, and A. Mukhin ... 76

Recognition of Convex Bodies by Tomographic Methods
V. Ohanyan ... 84

Validation of Risk Assessment Models for Breast and Ovarian Cancer–Related Gene Variants
W. Luther ... 89

Assessing Cancer and Stroke Risk Scoring Online Tools
N. Baloian, W. Luther, S. Peñafiel, and G. Zurita ... 106

Sales Goals Planning using Evidence Regression
N. Baloian, B. Panay, S. Peñafiel, J. A. Pino, J. Frez, and C. Fuenzalida ... 112

Information-Theoretic Investigation of Authenticated Steganographic Model in the Presence of Active Adversary

Mariam Haroutunian[0000-0002-9262-4173], Parandzem Hakobyan[0000-0002-5056-9591], Ashot Harutyunyan[0000-0003-2707-1039], and Arman Avetisyan[0000-0002-0434-2767]

Institute for Informatics and Automation Problems of NAS of RA
armar@sci.am, par_h@iiap.sci.am, aharutyunyan@vmware.com, armanavetisyan1997@gmail.com

Abstract. We study the information-theoretic model of stegosystem with active adversary. The legitimate sender as well as the adversary can be either active or passive, i.e. can embed or not a message in the sending data. The receiver's first task is to decide whether the communication is a covertext, data with no hidden message, or a stegotext, modified data with a hidden secret message. In case of stegotext, the second task is to decide whether the message was sent by a legitimate sender or from an adversary. For this purpose an authenticated encryption from the legitimate sender is considered.
Two-stage statistical hypothesis testing approach is suggested from the receivers point of view. In this paper a logarithmically asymptotically optimal testing for this model is suggested. As a result the functional dependence of reliabilities of the first and second kind of errors in both stages is constructed.

Keywords: Steganography · Information-theoretic security · Hypotheses testing · LAO tests · Error probability exponents (reliabilities) · Authentication.

1 Introduction

The aim of steganography is communicating messages by hiding them within other data thereby creating a covert channel. Various models with various tasks have been studied. We are interested in information-theoretic investigations.

In [1] an information-theoretic model with passive adversary (who has read-only access to the public channel) was considered. Another information-theoretic model with active attacks (where adversary can read and write a message over an insecure channel) was studied in [2].

In this paper we study the information-theoretic model of stegosystem with active attacks, where adversary is allowed to have access to a read and write public channel and able to analyze and modify data. The legitimate sender (Alice) as well as the adversary (Eve) can be either active or passive, i.e. can embed or not a message in the sending data. The receiver's (Bob) first task is to decide whether the received data X is a covertext C, data with no hidden message, or stegotext S, modified data with a hidden secret message M. In case of deciding that the obtained data is stegotext, Bob has the extraction function, and the second task for Bob is to decide whether the extracted message was sent by Alice or Eve. For this purpose an authenticated encryption of message m with secret key K is considered. Depending on applications this encryption except authentication can include also secrecy requirements of hidden message.

Covertext is generated by a source according to a distribution P_C, stegotext has a distribution P_S according to certain embedding function. The distribution of secret key we denote by P_K. We assume that Eve knows all these distributions. For the authenticated encryption Alice generates the encrypted message according to P_{MK} and Eve can generate a message with distribution $P_M P_K$.

We suggest two-stage statistical hypothesis testing approach. On the first stage Bob has to decide if the data was generated according to P_C or P_S. In the case when Bob decides that stegotext is obtained, after extracting the secret message, on the second stage Bob has to decide if the message was generated according to P_{MK} or $P_M P_K$.

In classical statistical hypothesis testing problem a statistician makes decision on which of the two proposed hypotheses H_1 and H_2 must be accepted based on data samples. This decision is made on the certain procedure which is called test. Due to randomness of the data the result of this decision may lead to two types of errors: the fist type is called the error for accepting H_2 when H_1 is true and the second type error for accepting H_1 when H_2 is true. In such problems the aim is to find such a test, that reduces both types of errors as much as possible. The complexity of the task is that the two types of errors are interconnected, when the one is reduced the other one can get increased.

According to [3] - [7] the problem is solved for the case of a tests sequence, where the probabilities of error decreased exponentially as 2^{-NE}, when the number of observations N tends to the infinity. The exponent of error probability E is called the *reliability*. In case of two hypotheses both reliabilities correspond-ing to two possible error probabilities could not increase simultaneously, it is an accepted way to fix the value of one of the reliabilities and try to make the tests sequence get the greatest value of the remaining reliability. Such a test is called *logarithmically asymptotically optimal* (LAO). The problem of multiple hypotheses LAO testing was investigated in [8], [9], [10].

In this paper two stage logarithmically asymptotically optimal testing of the described steganographic model is suggested. We study the functional dependence of reliabilities of the first and second kind of errors of optimal tests in both stages. The proof of the result for first stage is similar to the result suggested in [11], where the problem of logarithmically asymptotically optimal testing of statistical hypotheses for the steganography model with a passive adversary is solved by the method of types [12]. For the second stage the approach studied in [13] was useful.

2 Notations and Definitions

Here we present some necessary characteristics and results of information theory [14], [15]. We denote finite sets by script capitals. The cardinality of a set $\mathcal{X}$ is denoted as $|\mathcal{X}|$. We denote random variables (RV) by X, S, C, K, M. Probability distributions (PD) are denoted by Q, P, G, V.

Let PD of RV K and M be

$$P_K \triangleq \{P_K(k), \quad k \in \mathcal{K}\},$$

$$P_M \triangleq \{P_M(m), \quad m \in \mathcal{M}\},$$

and the joint PD of RVs M and K be

$$P_{MK} \triangleq \{P_{MK}(m,k), \ m \in \mathcal{M}, \ k \in \mathcal{K}\}.$$

The conditional PD of RV M for given K is denoted as follows:

$$V = \{V(m|k), \ m \in \mathcal{M}, \ k \in \mathcal{K}\},$$

The joint PD P_{MK} can be also written as:

$$P_{MK} \triangleq P_K V = \{P_K(k)V(m|k), k \ \in \mathcal{K}, \ m \in \mathcal{M}\}.$$

The space of all joint PDs on finite set $\mathcal{M} \times \mathcal{K}$ is denoted by

$$\mathcal{Q}(\mathcal{M} \times \mathcal{K}) \triangleq \{Q : Q = Q(m,k), m \in \mathcal{M}, \ k \in \mathcal{K}\}.$$

The Shannon entropy $H_P(X)$ of RV X with PD $P \triangleq \{P = P(x), \ x \in \mathcal{X}\}$ is:

$$H_P(X) \triangleq -\sum_{x \in \mathcal{X}} P(x) \log P(x).$$

Then the mutual information of RV M and K will be written as:

$$I(M;K) \triangleq \sum_{m \in \mathcal{M}, \ k \in \mathcal{K}} P_K(k)V(m|k) \log \frac{V(m|k)}{P_M(m)}$$

The joint entropy of RVs M and K is the following:

$$H_{P_{MK}}(M,K) \triangleq -\sum_{m \in \mathcal{M}, \ k \in \mathcal{K}} P_{MK}(m,k) \log P_{MK}(m,k).$$

The divergence (Kullback-Leibler information, or “distance”) of PDs $G \triangleq \{G = G(x), \ x \in \mathcal{X}\}$ and P on $\mathcal{X}$ is:

$$D(G||P) \triangleq \sum_{x \in \mathcal{X}} G(x) \log \frac{G(x)}{P(x)}.$$

The divergence of joint PDs Q and P_{MK} on $\mathcal{Q}(\mathcal{M}\times\mathcal{K})$ is:

$$D(Q||P_{MK}) \triangleq \sum_{m\in\mathcal{M},k\in\mathcal{K}} Q(m,k)\log\frac{Q(m,k)}{P_{MK}(m,k)}.$$

When RV M and K are independent, then

$$D(Q||P_{MK}) = D(Q||P_M P_K) = \sum_{m\in\mathcal{M},k\in\mathcal{K}} Q(m,k)\log\frac{Q(m,k)}{P_M(m)P_K(k)}.$$

In particular, the divergence of PDs P_{MK} and $P_M P_K$ is:

$$D(P_{MK}||P_M P_K) \triangleq \sum_{m\in\mathcal{M},k\in\mathcal{K}} P_{MK}(m,k)\log\frac{P_{MK}(m,k)}{P_M(m)P_K(k)}$$

$$= \sum_{m\in\mathcal{M},k\in\mathcal{K}} P_K(k)V(m|k)\log\frac{V(m|k)}{P_M(m)} = I(M;K).$$

For our investigations we use the method of types, one of the important technical tools in Shannon theory [12], [16].

3 Formulation of Results

First stage: At the first stage, from the received data $\mathbf{x}=(x_1,...,x_L)$, $\mathbf{x}\in\mathcal{X}^L$, Bob must decide whether it is a covertext or a stegotext. Hence, Bob must accept one of two hypotheses

$$H_1 : P = P_S \quad \{\text{data is a stegotext}\}$$

$$H_2 : P = P_C \quad \{\text{data is a covertext}\}$$

The procedure of decision making is a non-randomized test φ_L, which can be defined by partition of the set of possible messages $\mathcal{X}^L$ on two disjoint subsets $\mathcal{A}_i^L$, $i=\overline{1,2}$. The set $\mathcal{A}_i^L$, $i=\overline{1,2}$ contains all data $\mathbf{x}$ for which the hypothesis H_i is adopted.

The first kind error probability, which is the probability of the rejection of the correct hypothesis H_1 is the following:

$$\alpha_{2|1}(\varphi_L) = P_S^L(\mathcal{A}_2^L).$$

The second kind error probability, which is the probability of the erroneous acceptance of hypothesis H_1 is defined as follows:

$$\alpha_{1|2}(\varphi_L) = P_C^L(\mathcal{A}_1^L).$$

The error probability exponents, called "reliabilities" of the infinite sequence of tests φ, are defined respectively as follows:

$$E_{2|1}^I(\varphi) \triangleq \varliminf_{L\to\infty} -\frac{1}{L}\log\alpha_{2|1}(\varphi_L),$$

$$E^I_{1|2}(\varphi) \triangleq \lim_{L\to\infty} -\frac{1}{L}\log \alpha_{1|2}(\varphi_L).$$

As defined in [7] the sequence of tests φ^* is called logarithmically asymptotically optimal (LAO) if for given positive value of $E^I_{2|1}$ the maximum possible value is provided for $E^I_{1|2}$.

The procedure for creating an optimal decision rule is similar to [11]. The functional dependence of the reliabilities of the first and second kind of errors is given by the following theorem:

Theorem 1. *For given $E^I_{2|1} \in (0, D(P_C||P_S))$ there exists a LAO sequence of tests, the reliability $E^{*,I}_{1|2}$ of which is defined as follows:*

$$E^{*,I}_{1|2} = E^{*,I}_{1|2}(E^I_{2|1}) = \inf_{P:\ D(P||P_S)\leq E^I_{2|1}} D(P||P_C).$$

When $E^I_{2|1} \geq D(P_C||P_S)$, then $E^{*,I}_{1|2}$ is equal to 0 .

Thus, for a given reliability of incorrectly rejecting the stegotext, we get the maximal reliability of wrongly accepting the stegotext.

Comment 1: Unlike model considered in [1], [11], here Bob has no additional information about whether Alice is active or passive. Therefore, considered stegosystem should not be *perfectly secure*, because otherwise Bob cannot find out that he has received a covertext or a stegotext. Hence, we assume that for distributions P_C and P_S, $D(P_C||P_S) > 0$.

If at the first stage Bob accepts the hypothesis H_1, which means that he decides that the data is a stegotext, then he uses the extraction algorithm to get the hidden message $\mathbf{m} = (m_1, m_2, ..., m_N)$.

Second stage: After the extraction using key sequence $\mathbf{k} = (k_1, k_2, ..., k_N)$ Bob has to decide whether Eve or Alice sent him that message. So he moves on to the second stage of hypothesis testing:

$$H_1: \quad Q = P_{MK}(m,k) \qquad \{\text{there was no attack}\}$$

$$H_2: \quad Q = P_M(m)P_K(k) \quad \{\text{there was attack}\}$$

For this testing the test Φ_N is defined by partition of the set $(\mathcal{M}\times\mathcal{K})^N$ on two disjoint subsets $\mathcal{B}^N_l$, $l = \overline{1,2}$. The set $\mathcal{B}^N_1$ contains all data pairs $(\mathbf{m},\mathbf{k})$ for which the hypothesis H_1 is adopted, which in our context means that message $\mathbf{m}$ is sent from Alice. Correspondingly, the set $\mathcal{B}^N_2$ contains all pairs $(\mathbf{m},\mathbf{k})$ for which the hypothesis H_2 is adopted, i.e. Bob decides that message is sent from Eva.

The probabilities of errors of the first and second kind by analogy to the case of the first stage are defined as follows:

$$\alpha^{II}_{2|1}(\Phi_N) = P^N_{MK}(\mathcal{B}^N_2), \ \text{(the first kind error probability)}$$

$$\alpha_{1|2}^{II}(\Phi_N) = (P_M P_K)^N(\mathcal{B}_1^N), \text{ (the second kind error probability).}$$

The error probability exponents of the infinite sequence of tests Φ, are defined respectively as follows:

$$E_{i|j}^{II}(\Phi) \triangleq \underline{\lim}_{N\to\infty} -\frac{1}{N}\log \alpha_{i|j}^{II}(\Phi_N), \;\; i \neq j, \;\; i,j = \overline{1,2}.$$

The second kind error probability in Bob's decision essentially coincides with the probability of Eve's succeeding. Hence, the maximum value of $E_{1|2}^{II}$ guarantees that the attacker will fail.

As in the First Stage, for given positive value $E_{2|1}^{II}$ we constructed the LAO sequence of tests Φ^* and the dependence of maximal value $E_{1|2}^{II}$ from $E_{2|1}^{II}$ is provided in the following theorem:

Theorem 2. *For given $E_{2|1}^{II} \in (0, D(P_M P_K || P_{MK}))$ there exists a LAO sequence of tests, the reliability $E_{1|2}^{*,II}$ of which is defined as follows:*

$$E_{1|2}^{*,II}\left(E_{2|1}^{II}\right) = \inf_{Q:\, D(Q||P_{MK}) \leq E_{2|1}^{II}} D(Q||P_M P_K).$$

When $E_{2|1}^{II} \geq D(P_M P_K || P_{MK})$, then $E_{1|2}^{*,II}$ is equal to 0 .

Comment 2: For given $E_{2|1}^{II} \in (0, D(P_M P_K || P_{MK}))$ the following holds:

$$E_{1|2}^{*,II} \leq D(P_{MK} || P_M P_K) = I(M;K).$$

In the proof of Theorem 2, the optimal division of the set $(\mathcal{M} \times \mathcal{K})^N$ is constructed using sets of types. According to the properties of types, it is substantiated that this is the optimal division of the test, and at the same time, the dependence of reliabilities is established.

4 Conclusion and Future Work

Two-stage statistical hypothesis testing approach is suggested from the receivers point of view in the stegosystem with active adversary. The logarithmically asymptotically optimal testing for this model is suggested. As a result the functional dependence of reliabilities of the first and second kind of errors in both stages is constructed.

The practical implementaion and experimentation of the optimal testing result for steganography applications will be carried out in our future work.

5 Acknowledgment

The work was supported by the Science Committee of RA, in the frames of the research project № 21T-1B151.

References

1. Cachin, C.: An information-theoretic model for steganography. Information and Computation 192, 41–56 (2004)
2. Shikata, J., Matsumoto T.: Unconditionally secure steganography against active attacks. IEEE Transactions on Information Theory 54(6), 2690–2705 (2008)
3. Hoeffding, W.: Asymptotically optimal tests for multinomial distributions. The Annals of Mathematical Statistics 36, 369–401 (1965)
4. Csiszár, I., Longo, G.: On the error exponent for source coding and for testing simple statistical hypotheses. Studia Sc. Math. Hungarica 6, 181–191 (1971)
5. Blahut, R.: Hypothesis testing and information theory. IEEE Transactions on Information Theory 20(4), 405–417 (1974)
6. Longo G., Sgarro, A.: The error exponent for the testing of simple statistical hypotheses: A combinatorial approach. Journal of Combinatorics, Information and System Sciences. 5(1), 58-67 (1980)
7. Birgé, L.: Vitesses maximales de décroissance des erreurs et tests optimaux associés. Z. Wahrsch. verw. Gebiete 55, 261–273 (1981)
8. Haroutunian, E.: On asymptotically optimal criteria for Markov chains. The First World Congress of Bernoulli Society, (in Russian) 2(3), 153–156 (1989)
9. Haroutunian, E.: Logarithmically asymptotically optimal testing of multiple statistical hypotheses. Problems of Control and Information Theory 19(5-6), 413–421 (1990)
10. Haroutunian, E., Haroutunian, M., Harutyunyan, A.: Reliability Criteria in Information Theory and in Statistical Hypothesis Testing. Foundations and Trends in Communications and Information Theory 4(2–3) (2008)
11. Haroutunian, M., Haroutunian, E., Hakobyan, P., Mikayelyan, H.: Logarithmically asymptotically optimal testing of statistical hypotheses in steganography applications. In Proceedings 1st CODASSCA Workshop, Yerevan, pp. 157–163. Logos, Berlin (2018)
12. Csiszár, I.:"Method of types". IEEE Transactions on Information Theory 44(6), 2505–2523 (1998)
13. Maurer, U.: Authentication theory and hypothesis testing. IEEE Transactions on Information Theory 46(4), 1350–1356 (2000)
14. Blahut R.: Principles and Practice of Information Theory. Addison-Wesley, Reading, MA (1987)
15. Cover, T., Thomas, J.: Elements of Information Theory. Second Edition. Wiley, New York (2006)
16. Csiszár, I., Körner, J.: Information Theory: Coding Theorems for Discrete Memoryless Systems. Academic Press, New York (1981)

Another Approach for ZKRP Algorithms

Sergey Abrahamyan[1[0000-0002-8099-0205]]

Institute for Informatics and Automation Problems of NAS RA,
P. Sevak street 1, Yerevan 0014 Armenia
serj.abrahamyan@gmail.com

Keywords: Zero knowledge range proof · Order Preserving Encryption · Range Proof

Extended Abstract

During the last decade block chain technologies became more demanded and applicable. Development of blockchain technologies entail new cryptographic issues. One of them is a Zero knowledge range proof (ZKRP). ZKRP scheme allows to prove that a secret integer belongs to a certain interval without revealing any information about secret integer. For example in the payment system if the person X wants to transfer money to person Y then utilizing ZKRP one can prove that the transaction amount is positive, without revealing any information about X's money, otherwise such transaction would in fact transfer money in the opposite direction from Y to X.

The first ZKRP protocol was presented in 1995 by Damgard [6] and in 1997 by Fujisaki and Okamoto [7]. The first practical construction was proposed by Boudot in 2001 [3]. In 2016 Bünz et al [4] proposed a new idea of constructing ZKRP with a very small proof size which they called Bulletproof. The idea, similar to some other schemes, is to decompose the secret into the bit representation and using "Inner product proof" method prove that it belongs to the interval.

There is another interesting cryptographic primitive called Order Preserving Encryption-OPE (or Order Revealing Encryption-ORE). ORE is a deterministic symmetric encryption scheme the encryption algorithm of which produces ciphertexts that preserve numerical ordering of the plaintexts. OPE was proposed by Agrawal et al. [1] in 2004 as a tool to support efficient range queries on encrypted data. The first formal cryptographic treatment of OPE scheme were given by Boldereva et al. [2].

A number of OPE schemes have been proposed in recent years [1, 2, 10, 5, 8]. Unfortunately all these ORE schemes are not efficient to be used in practice. Concurrent with this works, Lewi and Wu [9] presented a new and efficient ORE scheme, which is based on the work of Chenette et.al. [5]. The ORE construction proposed by Lewi and Wu leaks less information about encrypted numbers, which is an important advantage.

In this paper the author proposes to transmute [9] into an efficient ZKRP scheme. As proposed scheme is based on [9] its short description is given below.

In [9] a large domain ORE scheme consists of three parts: Setup, Encryption(left, right) and Compare. Right encryption is used for encrypting values which are stored on server side. During the right encryption process for the each digit x_i of the value to be encrypted d digits (where d is the radix) of Z_3 are generated, which are comparison output of x_i and every element of radix. Then these numbers are permuted via permutation function. Thus each value is represented as a $d \times n$ table of elements of Z_3, where n is a maximal number of digits of upper endpoint. Each element of table is encrypted, and table stored in database. Left encryption is used only for making a search query. During the left encryption each digit of encrypting value is permuted via permutation function and encrypted. The server via the "Compare" algorithm compares left encrypted value with right encrypted value without revealing both of them.

The structure of "Large ORE Scheme" in [9] allows us to modify ORE scheme to ZKRP in the following manner: Encrypt the endpoints of a range interval via right encryption algorithm on verifier's side, encrypt the secret value via left encryption in prover side. Then prover sends the secret value to a verifier. The latter verifies if the secret value is smaller than upper endpoint and bigger than the lower endpoint. In order to provide completeness and soundness for new ZKRP it is important to design new key management system. Recall that completeness means that proof is accepted by the verifier if the statement or assertion is true. In other words, an honest verifier will always be convinced of a true statement by an honest prover. Soundness means that in the case of false fact the verifier rejects the proof, which indicates that a cheating prover can cheat an honest verifier with a negligible probability.

The other important property of ZKRP is a trusted setup. Many ZKRP constructions depend on a trusted party. A trusted party generates and provides necessary parameters for both prover and verifier. Some ZKRP algorithms avoid the trusted setup [8] which is an obvious advantage.

The proposed ZKRP scheme assures completeness and soundness and minimizes the dependence from trusted setup. The data concerning the algorithm's performance, security parameters and other details will be presented in the full paper to be prepared.

References

1. Agrawal, R., Kiernan, J., Srikant, R., Xu, Y.: Order-preserving encryption for numeric data. In: Proceedings of the ACM SIGMOD International Conference on Management of Data, pp. 563-574. Paris, France (2004).
2. Boldyreva, A., Chenette, N., Lee, Y., O'Neill, A.: Order-preserving symmetric encryption. In: Advances in Cryptology - EUROCRYPT 2009, pp. 224-241. Springer Berlin, Heidelberg (2009).
3. Boudot, F.: Efficient proofs that a committed number lies in an interval. In: Preneel,B. (ed.): Advances in cryptology - EUROCRYPT 2000, pp. 431- 44. Springer, Berlin (2000).
4. Bünz, B., Bootle, J., Boneh, D., Poelstra, A., Wuille, P., Maxwell, G.: Bulletproofs: Short proofs for confidential transactions and more. In: 2018 IEEE symposium on security and privacy (SP), pp. 315-334. IEEE, New York (2018).

5. Chenette, N., Lewi, K., Weis, S.A., Wu, D.J.: Practical order-revealing encryption with limited leakage. In: Fast Software Encryption - 23rd International Conference, FSE 2016, March 20-23, Revised Selected Papers, pp. 474-493. Bochum, Germany (2016). https://www.overleaf.com/project/6227a2f5d5c64893ad6209ac
6. Damgård, I.: Practical and provably secure release of a secret and exchange of signatures. J. Cryptol. 8(4), 201-222 (1995).
7. Fujisaki, E., Okamoto, T.: Statistical zero knowledge protocols to prove modular polynomial relations. In: Kaliski, B.S. (ed): Advances in cryptology - CRYPTO '97, pp. 16-30. Springer, Berlin (1997).
8. Lacharite, M., Minaud, B., Paterson, K. G.: Improved reconstruction attacks on encrypted data using range query leakage. In: 2018 IEEE Symposium on Security and Privacy (SP), pp. 297-314. IEEE, New York (2018).
9. Lewi, K., Wu, D. J.: Order-revealing encryption: New constructions, applications, and lower bounds. In: Proceedings of the 2016 ACM SIGSAC Conference on Computer and Communications Security, pp. 1167-1178. Vienna, Austria (2016).
10. Teranishi, I., Yung, M., Malkin, T.: Order-preserving encryption secure beyond one-wayness. In: Advances in Cryptology - ASIACRYPT 2014, pp. 42-61. Springer, Berlin, Heidelberg (2014).

Preserving Location and Query Privacy Using a Broadcasting LBS

Pablo Torres-Osses[1][0000−0003−3467−0945], Patricio Galdames[1,2][0000−0003−3051−2413], Claudio Gutierrez-Soto[1,2][0000−0002−7704−6141], and Geraldine Solar[1][0000−0001−6856−7933]

[1] Universidad del Bío-Bío, Concepción 4051385, Chile
{paitorre,geraldine.solar1701}@egresados.ubiobio.cl
[2] Group of Smart Industries and Complex Systems (gISCOM)
Universidad del Bío-Bío, Concepción 4051385, Chile
{pgaldames,cogutier}@ubiobio.cl

Abstract. Untrustworthy LBS can compromise their clients by releasing their location and sensitive query attributes without permission. The traditional approach to protect users' privacy is building a set of dummy queries that aim to protect location using *K-anonymity* and the query attributes, using *l-diversity*. Here, privacy is achieved at the expense of the LBS server since it can quickly become a bottleneck when the number of LBS requests is increased. This work proposes that the LBS server periodically broadcasts selected public data into a public air channel. Meanwhile, users intend to solve their queries by listening to this channel. In this way, privacy is preserved without increasing the server workload. However, when a user does not find the needed data in the air, it submits a location cloaked query (LCQ) to the LBS server. We propose some approaches on how to adjust a periodic data broadcasting program that balances the response time incurred in locating data on the air channel and the loss of privacy when users directly contact the LBS. Thus, our idea is that the LBS server treats the wireless channel as a cache memory and keeps broadcasting the most relevant data periodically.

Keywords: Query Privacy; Location Privacy; Broadcasting LBS.

1 Introduction

The last decade has witnessed a breathtaking revolution in the mobile devices industry. Advances in microelectronics and wireless technologies have resulted in significant improvements in data processing, storage, communication and other, allowing smart devices to execute ever-increasingly complex applications [19]. In addition, new smart devices have experienced important reductions in size and cost. That, along with the integration of other features (such as sensors and others), fostered a new generation of software applications and systems, which greatly impact the way people act. Today, mobile technologies are massively used and have become part of people's daily life.

Lately, Location-Based Service (LBS) has gained significant attention due to its wide application areas. LBS is possible because of the availability of integrated Global Positioning System (GPS) modules. An LBS is a software service used to provide information or other usage based on the device location [24]. Examples of LBS systems include navigation software, location-based games, ride service hailing apps, touristic details about a city and others. Nevertheless, the large-scale use of LBS involves security and anonymity issues. Indeed, the location itself has become a user's quasi-identifier and, therefore, it can allow the LBS to determine a user's identity [5]. Moreover, user positioning can reveal behavioral patterns, such as frequent schedules and routes from work to home. This situation is even worst when location information is correlated with sensitive attributes of the same query since LBS can conclude more private details about their customers. This situation should not be an issue if the LBS would protect their clients' information adequately. Many examples about the leak of private information into the public can be found in the press [4].

Several techniques have been proposed to protect a user's location and sensitive query attributes when accessing a LBS. For protecting location privacy, a first approach applies spatial transformations [12], and query processing is performed using secure multiparty computation. Others are based on location disturbance [31, 3], which consists of perturbation techniques that add some (controlled) random noise to the user's exact location. Another trend based on the reduction of location resolution [9, 20, 29, 2, 7, 10], which is the most widely studied. This last one seeks to build a cover-up region that relies on the concept of k-anonymity [26]. K-anonymity looks for finding a geographical area, called a cloaking region, that contains different locations, one of which is the user's actual location (the other sites correspond to other possible positions where the user could be). This technique is complemented with l-diversity [18]. Here, sensitive attributes of a specific location-based query are indistinguishable from other $l-1$ other queries located in the same position.

Here, privacy protection is achieved at the expense of the LBS because for answering a single query, the LBS must answer many other dummy queries increasing dramatically its workload and affecting its efficiency when it faces many requests. To address the workload and privacy issues, some authors [16, 7] assume that a third reliable party called as *the anonymity server*, performs proactive public data broadcasting. Here, users intend to fetch query answers by listening to a public air channel in which data is broadcast. When data is not found in the channel, users need to access the LBS directly. To alleviate the server's workload, in [16], users submit a traditional Location-based query putting into risk their privacy. On the contrary in [7], users protects their queries using k-anonymity techniques but affecting the server's workload. None of the aforementioned works study the effects of the size of the data broadcast. If this size is large, there is a high probability of finding the needed data within the air channel (and protecting privacy as well), however the response time becomes larger.

This article considers a location and query privacy-aware LBS server that wants to proactively broadcast location-based data periodically to users moving in a specific geographic area. This research wants to find a proper balance between privacy and response time when users change their needs. For example, assume tourists, located at noon at a public square, desire restaurant information. However, these tourists could be interested in visiting attractions like a park or museum after some hours. Therefore the air channel must be updated with fresh and the most needed data.

Our idea is to view the air channel as a "cache memory" from which users listen to the most needed data before directly accessing the "main memory" which is the LBS server. However, choosing what data to broadcast is challenging because many queries are dummy, and therefore, their corresponding answers could be useless. Moreover, the LBS does not know the number of queries answered by the air channel since the LBS receives only those queries whose answers were not found "on the air". Our key idea is to use these latter queries to estimate our air channel's success. We will discuss a few approaches to do so.

The remainder of this paper organizes as follows: In Section 2, Related work is presented. In Section 3, a system overview is exposed. Section 4 discusses our ideas to implement a cache memory over the air channel. Finally, Section 5 provides some perspectives on future work and the conclusions.

2 Related Work

In this paper, we present the related work based on two categories. In the first one, we expose approaches that perform data broadcasting to preserve location privacy. The second category, where privacy is not a concern (all data is valid), summarizes some relevant approaches to selecting what data should be broadcast first on an air channel.

2.1 Location-privacy aware Query Processing

In [15], the researchers provide a three-tier architecture in which the anonymizer server takes the role of a broadcast server. Here, users submit their queries to the anonymizer to protect their location privacy which subsequently forwards them to the LBS. Query results are publicly broadcast only once to all mobile users, and data broadcast consists of query answers being currently asked to the LBS server. Our proposal distinguishes itself from this work in two aspects. First, our broadcast server periodically repeats data broadcast to everyone even when no queries have been received at the LBS. The second one is that we consider that the broadcast combines some selected current query answers with a few past transmitted ones.

Galdames et al. [7] propose a similar three-tier architecture as the one proposed by [15] that aims to organize query answers considering the followings four performance metrics; server's throughput, user's response time, effectiveness, and unfairness of the broadcast. However, the main drawbacks of their techniques are; first, it does not limit the size of a broadcast, and in the worst-case scenario, the LBS can broadcast the entire database. The second one arises when users contact the LBS server, they do not get their answers immediately but in the next broadcast. All these situations can negatively affect the response time. The third one is that this work does not provide a metric to measure the privacy loss when users access an LBS. Finally, our work aims to extend the notion of LCQs to protect query privacy, which is another dimension of privacy not considered by [7, 15].

Li et al. [14] propose a suite of privacy-preserving Location Query Protocol based on cryptography. This work assumes that POIs correspond to sensitive information (like other users' current location), and they are only transmitted to the querier. On the contrary, in our work, we assume POIs correspond to public locations like a restaurant, a museum or a hospital, and answers can be publicly known to everyone. This way, users can eventually find answers on the air channel without revealing their presence to the LBS. Similarly, Schlegel et al. [25] proposes a system for protecting location privacy based on a semi-trusted third party using cryptography as well. Here, answers are only provided to the querier, although other users could be interested in the same data. Although cryptography will become the basis for protecting privacy in the future, current solutions incur higher communication overhead and higher server workload.

2.2 Scheduling techniques for precise on-demand queries

This subsection includes related works, proposing different techniques to sort in real-time data and then broadcasting it publicly to everyone. In all this work is assumed that users release their exact location and query to the LBS server, and therefore, privacy is not a concern.

Request based on single-item Aksoy and Franklin [1] propose a scheduling algorithm, named RxW, for large-scale on-demand data broadcast. RxW corresponds to a metric where R is the number of answered requests, and W is the maximum request time which has not been processed. The authors proposed three heuristics; the first one exhaustively looks for the broadcast with maximum RxW, the second one prunes the search space, and the third one makes up the scalability in favour of the response time for the average and worst-case.

In [28], the researchers propose a scheduling algorithm named SIN-α, which considers the urgency and productivity of serving sending requests. The scheduling algorithm has the following characteristics: First, data items answering pending requests have the highest priority to be transmitted. Second, the algorithm considers the deadline associated with each query to determine its order in the data transmission. Third, it can only join a new request with a pending one if both requests ask for the same data item. Fourth, the data elements have a fixed size.

Request based on multi-item In [6], six single-item request-based scheduling algorithms are analyzed in time-critical multi-item request environments.

Scheduling algorithms such as SIN-α, RxW y LWF impair their performance when dealing with the multi-item request.

Wang [27] explores five scheduling algorithms that intend to minimize the worst access time for both popular and unpopular queries. These algorithms aim to find the set of answers as large as possible so that its worst access time is the minimum possible. According to the authors, this challenge can be solved optimally if mobile users' access patterns are present in some particular form.

Lu et al. [17] analyze the multi-item request scheduling on-demand wireless data broadcasts to minimize the average access latency. The authors developed a two-stage scheduling scheme to sort the requested data items. The first stage implies the selection of data items to carry out the broadcast in the next period. The second stage involves scheduling the broadcasting order for the data items chosen in the first stage. This work proposes two scheduling algorithms named MTRS (Maximum Throughput Request Selection) and MLRO (Minimum Latency, Request Ordering).

It is essential to highlight that our work can be classified under the multi-item scheduling techniques.

3 System Overview

In this section, we first present some definitions, then the system architecture, and finally, how the query results are organized to be transmitted in an air channel.

3.1 Preliminary Details

There are a variety of Location-Based Queries, but the most popular ones are the K-Nearest Neighbor Queries and Range Queries. In [7], the authors show that a K-NN query can be converted to a circular range query centred at the query owner's position, and radius equals the distance between this owner and the location of its K-th nearest neighbour. Accordingly, we assume that all user requests are range queries. Hence, we understand a range query as follows:

- *Range query (rq).* A user is looking for all Points of Interest (POIs) whose location falls into a given square region and satisfying a given condition. For simplicity, we assume this condition is a single description of the type of POI that the user is looking for. This type could be restaurant, hotel, hospital, ATM, and so on. The square region is defined by the client's position and size (r), which its owner defines. Thus, we say *the answer of a range query rq* corresponds to all POIs of a given type whose locations are within the square area defined by the query.

Additionally, in the research literature, the K-anonymity approach has been extended to a *spatial K-anonymity*. Here, a region called *the cloaking region* is built for each user to protect its location privacy. A few authors have defined a *cloaking region* as the set of positions that contains the exact user's location, and at least other $K-1$ users' locations [9, 30]. Other researchers assume a cloaking region is a single and convex area enclosing the user's exact location and also the locations of others [21]. Finally, other authors consider a cloaking region as a list of fragmented regions [13, 23, 22, 8] and the $K-1$ positions are chosen in such a way that the target user is highly likely to be located there as well. Although our techniques can work with any of these approaches, we assume the latter one in this work.

- *Cloaking region (R).* A set of K disjoint spatial locations in which a user can be with high probability. The value K is the anonymity degree a user demands, and one location of R must be the real user's position.

For simplicity, we assume that the entire network area is split into LxL cells, and a user's cloaking region consists of a set of K different cells [23, 8], and one of these cells contains the exact location of the user. Since we have introduced the definition of a range query and a cloaking region, now we can introduce a Location-Cloaked Query (LCQ):

- *Location-Cloaked Query (q).* Given a range query rq and a cloaking region R, we define an LCQ as the set of all K range queries whose shape is the same as rq, but each query is anchored at each location of R. Thus, answering an LCQ means answering each of its corresponding range queries.

However, as many authors have shown, k-anonymity is not enough to preserve privacy [18] and proposes the concept of l-diversity. A location-based query is grouped with other nearby queries but has at least l-distinct sensitive attributes. In other words, all these queries must all be of distinct type of POI. For example, two users in the same neighbourhood demand 2-diversity. The first is looking for a nearby hospital, and the second is looking for a liquor store. These two queries have different semantics and can be considered sensible queries. Therefore the system can submit them as one big query to reduce the loss of query privacy since the LBS cannot distinguish which one comes from a specific user.

- *A set of dummy queries (D(lcq)).* Given an LCQ lcq, we define a set of dummy queries as a collection of LCQs having at least l-distinct type of POIs.

3.2 A Broadcasting LBS

Without prejudice to the generality, we assume there are many mobile GPS-capable users and a single LBS server, which is responsible for handling all data related to the point-of-interests (POIs). Each POI consists of a geographic position and any description details about this position. The LBS indexes all POIs for fast retrieval of the description from a database using a simple approach proposed by [7, 15]. This approach partitions the network domain into a set of cells, as shown in figure 1. The triangles represent some type of POI.

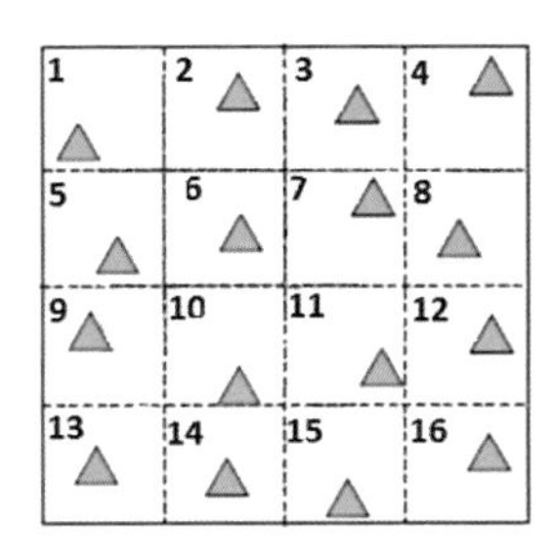

Fig. 1. Grid Partition of the application domain

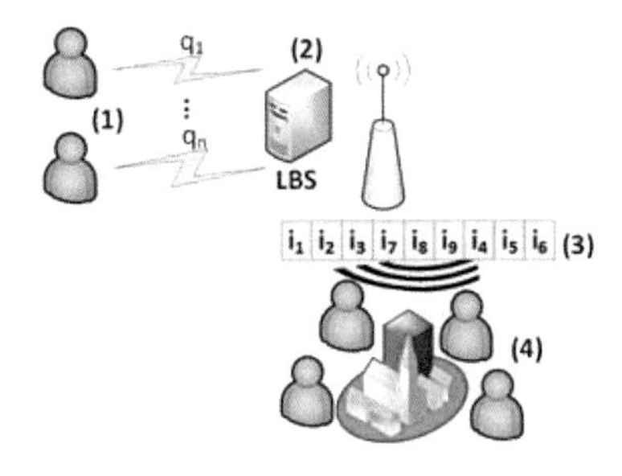

Fig. 2. System Architecture

If the number of POIs located inside a cell exceeds some threshold, the LBS recursively partitions this cell. The LBS server can maintain in the main memory the indexing nodes, which stores the cell partitioning hierarchy. On the other hand, all data nodes, each keeping a list of POIs, are stored on a disk and retrieved when needed. In this paper, we assume the cost of processing an LCQ is proportional to the number of data items that the server needs to recover from the database.

The LBS server periodically broadcasts data related to all POIs within a chosen list of cells to all users moving in the network area. For simplicity, we assume the LBS server has a permanent public air channel to disseminate data to everyone. Also, the LBS server can open a private air channel to transmit data back to a user for a short period.

The system architecture is displayed in figure 2. When users do not find answers to their queries, they submit them directly to the LBS at step (1). The LBS processes them at step (2), and sends them back to each user its corresponding answer through a private air channel. In parallel, the LBS also chooses, sorts and broadcasts new data in tandem throughout the public air channel at step (3). Finally, new users listen to the public air channel (4) and download the data required by them without directly accessing the LBS server.

Each mobile user knows the network partitioning but has no details/description of the nearby POIs. Before submitting a range query to the LBS, each user listens to the air channel until it listens to a unique mark that indicates the end of the existing data transmission. If a user finds all POIs within its range query satisfying a given condition or type, we say it satisfies its query, and this user does not need to contact the LBS. On the contrary, if it does not find the data required to answer its query, it needs to create a set of location cloaked queries. First, it requests a cloaking region to protect location privacy and also a set of dummy queries to protect query privacy to a third trusted party. Finally, the user submits all these LCQs to the LBS.

For simplicity, we assume the LBS server keeps a query queue Q, and all incoming queries are first placed in this queue. After that, the server has completed the transmission of some data; it processes in batch all outstanding queries in Q and also transmits each specific answer to its corresponding user throughout private air channels. Our problem arises when the LBS server needs to decide what data should be broadcast to everyone. For one side, the server can reduce its workload in query processing by transmitting a large amount of data. For the other side, the server must limit the amount of transmitted data to reduce the users' response time.

To facilitate our analysis, we define the following key concepts:

- *Server Workload.* The number of queries answered by the LBS server in some period throughout private channels. The higher this value is, the higher the probability the server becomes a bottleneck.
- *Response Time.* The time elapsed since a user sent an LCQ to the LBS server until it got the last data item that answered its query.
- *Usefulness.* Let d be the total amount of data that a client downloads from the LBS server and d' be the amount of data needed for its interest. Then, client's download usefulness (U) is defined as $\frac{d'}{d}$.A higher U means less client battery power is consumed in listening to irrelevant data (due to location cloaking).
- *Effectiveness.* Let d'' be the amount of data needed for all users during some period T and let B and ordered set containing the amount of data transmitted in the air channel during T. Then, the effectiveness (E) of the public broadcast is defined as $\frac{d''}{|B|}$. A higher E means the public broadcast contains relevant data to their clients.

.

Our idea to tackle our problem is to consider the broadcast channel as a cache memory. Usually, the performance of the cache memory is determined by its hit ratio. For operating systems, every time data is found within the cache, it is an indication this data is relevant. However, our scenario is more challenging because the LBS does not know with certainty what data is really needed and provides more data than is really expected.

4 Air Channel Policies for a Broadcast LBS

Let Q be the set of outstanding LCQs at some time. When Q is processed, the LBS server obtains the list C of all grid cells containing POIs relevant for all queries in Q. For example, consider figure 3, in which two LCQs, denoted as Q_1 and Q_2, need to be processed. The server retrieves all cells overlapping the query which corresponds for Q_1, the cells $\{5, 6, 7, 9, 10, 11, 13, 14, 15\}$ and for Q_2 are the cells $\{2, 3, 4, 6, 7, 8, 10, 11, 12\}$. In this figure, the black dot represents a user's location, and the red square identifies the cell in which a user is currently moving. Finally, the server retrieves all data about the POIs in these cells from its database.

When all relevant cells and their POIs are retrieved, the first server's goal is to establish an order of transmission of the POIs in the public air channel. We

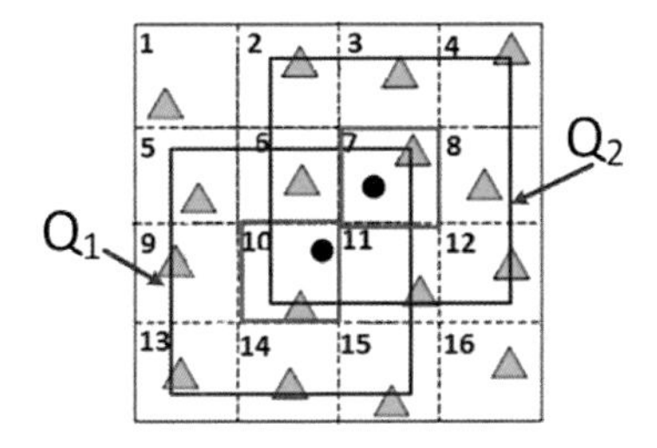

Fig. 3. Example of the processing of two LCQs

define B as a FIFO queue that establishes the order of transmission of the POIs related to C. It is relevant to mention that if the size of C is larger than the maximum size allowed for B, denoted as $|B_{max}|$, only a subset of C is broadcast to everyone throughout the public air channel.

The algorithm 1 establishes a broadcast program, which indicates what data should be transmitted first in the public air channel.

Algorithm 1: Preparing a Broadcast Program

Data: Q the list of outstanding queries at some point in time.
Result: A broadcast program B.

1 $B \leftarrow \emptyset$;
2 $Q_{temp} \leftarrow Q$;
3 **while** $Q_{temp} \neq \emptyset$ **do**
4 $\quad Q_{eff} \leftarrow Q_{temp}$;
5 $\quad$ **while** $Q_{eff} \neq \emptyset$ **do**
6 $\quad\quad C_{eff} \leftarrow$ *all cells requested by queries in* Q_{eff};
7 $\quad\quad c \leftarrow$ *cell with the highest priority in* C_{eff} *and* $c \notin B$;
8 $\quad\quad$ *B.push(c);*
9 $\quad\quad$ *Remove from* Q_{eff} *and* Q_{temp} *all queries whose relevant cells are all included in* B;
10 $\quad$ **end**
11 $\quad$ **while** $|B| > |B_{max}|$ **do**
12 $\quad\quad$ *Remove the cell with lowest priority from* B;
13 $\quad$ **end**
14 **end**
15 *Return* B

Lines 5-10 choose the highest priority cell, and Q_{eff} stores all queries overlapping such a selected cell. It is essential to highlight that a query response is composed of multiple cells or items, and according to [7], to achieve an average low response time is necessary to transmit all cells relevant to the query as close in as possible in the broadcast program. Lines 11-13 filter out those cells with low priority only when the size of the broadcast program is larger than the maximum one sets for the system.

One final aspect to discuss is how the broadcast program (aka cache memory) is adjusted when new queries arrive at the LBS server. Here we have several ideas based on the following definitions:

- *Popularity of a cell.* We calculate this priority as the number of queries overlapping a cell. As suggested by [7], the more queries intersect a cell, the higher the probability some user needs this data from that cell. Let us consider figure 3; we can say cell 7 is more popular than cell 5 since two queries overlaps cell 7, but only one for cell 5.
- *Popularity of a query.* The average popularity of the cells overlapped by this query divided by the number of cells overlapped by the query. We want to penalize those queries having a large cloaking region since this situation will make the LBS transmit useless data.

Select data depending on the the popularity of their corresponding cells, the time a cell has been present in a broadcast program, and the number of current queries that are answered when a chosen cell is included in the broadcast program. Given the size of the broadcast program (aka size of our cache), our idea is to select a few specific cells from those requested in the current answered queries and cells chosen in the latest broadcast program. In the best scenario, the LBS does not receive new queries, which means it keeps broadcasting the latest computed broadcast program; otherwise, the latest program is adjusted.

5 Conclusion

In this work, we have considered a location-privacy aware Location-Based Service (LBS) that works at the application layer and proactively broadcasts location-based items to all users in a service area. An item consists of the location of a Point of Interest and extra information related to this point. Usually, a user demands to know all POIs within a particular region, and thus a correct answer consists of many items. When a user needs a service, it first listens to the data being currently broadcast and usually tries to find multiple items. However, when relevant data is not found for all needed cells, it submits a Location-Cloaked Query to the LBS.

In this paper, we have discusses these challenges of periodically broadcasting location-based items in a way that balances users' response time, users' location & query privacy leaks, and the server's workload. We expect to propose a system that regularly adjusts a broadcast program based on current pending outstanding LCQs and previous computed programs. The challenge is to predict which data are not needy and can be removed for future broadcast. Finally, we plan to test the impact of inserting some air index in a broadcast program as suggest by [11].

6 Acknowledgments

This work supported by the Universidad del Bío-Bío of Chile and the Group of Smart Industries and Complex Systems (gISCOM) under grant DIUBB 195212 GI/EF.

References

1. Aksoy, D., Franklin, M.: RxW: a scheduling approach for large-scale on-demand data broadcast. ACM/IEEE Trans. on Networking **7**(6), 846–860 (1999)
2. Alabdulatif, A., Kumarage, H., Khalil, I., Yi, X.: Privacy-preserving anomaly detection in cloud with lightweight homomorphic encryption. Journal of Computer and System Sciences **90**, 28–45 (2017)
3. Andres, M., Bordenabe, N., Hatzikokolakis, K., Palamidessi, C.: Geo-indistinguishability: differential privacy for location-based systems. In: Proceedings of the 2013 ACM SIGSAC conference on Computer & communications security (CCS'13). pp. 901–914 (Nov 2013)
4. is Beautiful, I.: World's Biggest Data Breaches &Hacks. https://www.informationisbeautiful.net/visualizations/worlds-biggest-data-breaches-hacks, updated Oct. 2021
5. C., B., S., M., X.S., W., D., F., S, J.: Anonymity and historical-anonymity in location-based services. Privacy in Location-Based Applications. Lecture Notes in Computer Science **5599**, 1–30 (2009)
6. Chen, J., Lee, V.C., Liu, K.: On the performance of real-time multi-item request scheduling in data broadcast environments. Journal of Systems and Software **83**(8), 1337–1345 (2010)
7. Galdames, P., Cai, Y.: Efficient processing of location-cloaked queries. In: Proc. of IEEE Int'l Conf. on Computer Communications (INFOCOM'12). pp. 2480–2488 (2012)
8. Galdames, P., Gutiérrez-Soto, C., Curiel, A.: Batching location cloaking techniques for location privacy and safety protection. Mobile Information Systems pp. 9086062:1–9086062:11 (2019)
9. Gruteser, M., Grunwald, D.: Anonymous Usage of Location-based Services through Spatial and Temporal Cloaking. In: Proc. of the Int'l ACM Conf. on Mobile Systems, Applications and Services (MobiSys'03). pp. 31–42 (2003)
10. Gutiérrez-Soto, C., Galdames, P., Faúndez, C., Durán-Faúndez, C.: Location-query-privacy and safety cloaking schemes for continuous location-based services. Mobile Information Systems, Hindawi (2022)
11. Hu, Q., Lee, W.C., Lee, D.: Indexing techniques for wireless data broadcast under data clustering and scheduling. In: Proceedings of the eighth international conference on Information and knowledge management (CIKM'99). pp. 351–358 (1999)
12. Indyk, P., Woodruff, D.: Polylogarithmic private approximations and efficient matching. In: Int'l Conf. on Theory of Cryptography (TCC'06). pp. 245–264 (Oct 2006)
13. Li, T.C., Zhu, W.T.: Protecting user anonymity in location-based services with fragmented cloaking region. In: 2012 IEEE International Conference on Computer Science and Automation Engineering (CSAE). pp. 227–231 (2012)
14. Li, X.Y., Jung, T.: Search me if you can: Privacy-preserving location query service. In: Proceedings of the IEEE INFOCOM. pp. 2760–2768 (2013)

15. Liu, F., Hamza-Lup, G., Hua, K.: Using broadcast to protect user privacy in location-based applications. In: Proc. of IEEE Globecom 2010 Workshop on Web and Pervasive Security (WPS 2010). pp. 1561–1565 (December 6-10 2010)
16. Liu, F., Hua, K., , Do, T.: A p2p technique for continuous knearest-neighbor query in road networks. In: Proc. of the 17th International Conference on Database and Expert Systems Applications (DEXA'07). pp. 264–276 (September 04 - 08 2007)
17. Lu, Z., Wu, W., Li, W.W., Pan, M.: Efficient scheduling algorithms for on-demand wireless data broadcast. In: IEEE INFOCOM 2016-The 35th Annual IEEE International Conference on Computer Communications. pp. 1–9 (2016)
18. Machanavajjhala, A., Gehrke, J., Kifer, D., Venkitasubramaniam, M.: L-diversity: privacy beyond k-anonymity,. In: 22nd International Conference on Data Engineering (ICDE'06). pp. 24–24 (2006)
19. Mijic, D., Draškovic, D., Varga, E.: Scalable Architecture for the Internet of Things. O'Reilly Media, Inc., 1 edn. (2018)
20. Mokbel, M., Chow, C., Aref, W.: The new casper: Query processing for location services without compromising privacy. In: Proc. of ACM Int'l Conf. on Very Large Databases (VLDB'06). pp. 763–774 (September 12-15 2006)
21. Molina-Martínez, C., Galdames, P., Duran-Faundez, C.: A distance bounding protocol for location-cloaked applications. Sensors **18**(5), 1337 (2018)
22. Niu, B., Gao, S., Li, F., Li, H., Lu, Z.: Protection of location privacy in continuous lbss against adversaries with background information. In: 2016 Int'l Conf. on Computing, Networking and Communications (ICNC). pp. 1–6 (Feb 2016)
23. Niu, B., Li, Q., Zhu, X., Cao, G., Li, H.: Achieving k-anonymity in privacy-aware location-based services. In: Proc. of The 33rd Conference on Computer Communications (INFOCOM'14). pp. 754–762 (April 27- May 2 2014)
24. Schiller, J., Voisard, A.: Location-Based Services. Elsevier, 1 edn. (2004)
25. Schlegel, R., Chow, C.Y., Q. Huang, Q., Wong, D.: User-defined privacy grid system for continuous location-based services. IEEE Transactions on Mobile Computing **14**(10), 2158–2172 (2015)
26. Sweeney, L.: k-anonymity: a model for protecting privacy. International Journal on Uncertainty,Fuzziness and Knowledge-based Systems **10**(5), 557–570 (2002)
27. Wang, J.Y.: Set-based broadcast scheduling for minimizing the worst access time of multiple data items in wireless environments. Information Sciences **199**, 93–108 (2012)
28. Xu, J., Tang, X., Lee, W.C.: On scheduling time-critical on-demand broadcast. IEEE Transactions on Parallel and Distributed Systems **17**(1), 3–14 (2006)
29. Xu, T., Cai, Y.: Feeling-based Location Privacy Protection for Location-based Services. In: ACM Conference on Computer and Communications Security (CCS'09). pp. 348–357 (November 2009)
30. Xu, T., Cai, Y.: Feeling-based location privacy protection in location-based services. In: Proc. of ACM Int'l Conf. on Computer and Communications Security (CCS'09). pp. 348–357 (September 12-15 2009)
31. Zhuang, Y., Yang, A., Hancke, G., Wong, D., Yang, G.: Energy-efficient distance-bounding with residual charge computation. IEEE Transactions on Emerging Topics in Computing **8**(2), 365–376 (Oct 2017)

On Diagnosing Cloud Applications with Explainable AI

Ashot N. Harutyunyan[1,2,3[0000-0003-2707-1039]], Nelli Aghajanyan[1[0000-0002-3560-7502]], Lilit Harutyunyan[1[0000-0002-9558-9385]], Arnak Poghosyan[1,4[0000-0002-6037-4851]], Tigran Bunarjyan[1[0000-0003-4427-4284]], and A.J. Han Vinck[5,6[0000-0003-3437-3676]]

[1] VMware Eastern Europe, 0014 Yerevan, Armenia
{aharutyunyan;aghajanyann;lharutyunyan;
apoghosyan;tbunarjyan}@vmware.com
[2] AI Lab at Yerevan State University, 0025 Yerevan, Armenia
[3] Institute for Informatics and Automation Problems of NAS RA, 0014 Yerevan, Armenia
[4] Institute of Mathematics of NAS RA, 0019 Yerevan, Armenia
[5] Institute of Digital Signal Processing, University of Duisburg-Essen,
47057 Duisburg, Germany
[6] University of Johannesburg, PO Box 524, Auckland Park 2006, South Africa
han.vinck@uni-due.de

Abstract. Machine learning (ML) methods and solutions are of high importance for real-time automated identification and remediation of potential sources of performance degradations in cloud operations to minimize their impacts. At the same time, natural lack of human-annotated or labeled data in this domain is a serious bottleneck for reliable and automated root cause analysis (RCA). We propose ML approaches that address this problem from different perspectives. Specifically, we describe ideas and methods for predictive modeling of application components subject to a Key Performance Indicator (KPI), including a data self-labeling technique to benefit from supervised learning algorithms. For an exemplary distributed application and its selected KPIs such as a latency metric, interpretable models are trained, validated against expertise of application developers, and used for producing run-time root-cause recommendations on KPI abnormalities.

Keywords: Cloud infrastructures, Key Performance Indicators (KPI), automated root cause analysis, explainable ML/AI.

1 Intelligent Diagnosis of Cloud Environments

1.1 Explaining Factors of Misbehaviors

Expert efforts and knowledge are not anymore adequate for reliable management and quick remediation of misbehaving components of modern cloud environments tending to build self-driving capabilities (such as an application KPI optimization). Backtracking and finding root cause of failures in those distributed environments with high degree of sophisticated interrelations among data center objects is an extremely hard problem. Machine learning helps to automate the management of such complex systems [1] that contain thousands of objects like virtual machines (VM), Hosts, datastores, via monitoring millions of time series metrics, huge volume of logs, and application traces, to capture a high-resolution image of the entire stack.

Automated RCA with machine intelligence is a core problem in the self-driving data centers context. However, for gaining user trust in ML solutions it is also essential and preferable to build such technologies on top of interpretable models [2], [3]. There are multiple factors that hinder designing effective RCA solutions with ML for cloud computing infrastructures and applications, the main one is the lack or absence of labeled data. Operator or expert verified/annotated/labeled data sets are hard to obtain in this domain and ungeneralizable from one environment to another because of eco-system specifics.

In that context, we outline ideas and a prototype solution for automated RCA in terms of diagnosing KPI degradations (i.e., value range which is unacceptable or abnormal for the user) that target troubleshooting customer data centers, as well as self-diagnostics of the operations management products performing surveillance of those environments. The latter is enabling a proactive/predictive support capability of cloud management solutions, while collecting high-frequency telemetry data from the products.

Our proposal is built on the following ideas:

- Applying KPI metric as a source for generating labels for the entire data set of the application, while quantizing it into two or more class IDs. A parameter which can be a hard threshold on the KPI behavior required by the user, or, a high-quantile value of the same metric, splits this time series into two ranges where the application state is categorized as either normal or abnormal.
- Training regression and classification models to leverage those in predicting KPI abnormalities, while also evaluating relative variable/feature importance scores of those models to be employed for interpretability purposes.
- Applying decision trees and rule induction (a form of explainable AI) algorithms to derive consistent conditions on behaviors of the application features for recommending potential causes of abnormal KPI instances.

Overall, the objective of such a study is to identify important features and conditions of cloud applications subject to impact on KPIs. Based on this, intelligent cloud management solutions can provide recommender systems for optimizing applications performance (while indicating those important variables to be tuned) and predicting patterns causing unacceptable performance states (hence, accelerate the system recovery). This work focuses on experimental evaluation of the self-diagnostic use case of the technology leading cloud management solution vRealize Operations [4]. Fig. 1 depicts this product application in its functions to monitor and guard multi-cloud infrastructures. The diagram reflects three cloud environments built on 1) VMware compute/storage/network virtualization solutions vSphere, vSAN, and NSX [5], 2) such an infrastructure hosted in Microsoft Azure, and 3) native Azure service, respectively.

While this distributed application is intelligently managing various cloud environments, the product itself might greatly benefit from self-healing capabilities or automated recommendations for performance improvements and recovery from misbehaviors. Architectural specifics and variety of workload patterns at different types of clouds may affect/stress vRealize Operations performance differently. Therefore, for self-diagnostics purposes, special models need to be trained for each case with specific requirements on KPIs behavior set by users. Various ML algorithms are employed for comparative analysis including neural networks. Expert validation of discovered patterns promises wider adoptability of the approaches in real world scenarios with limited or unavailable annotated data sets.

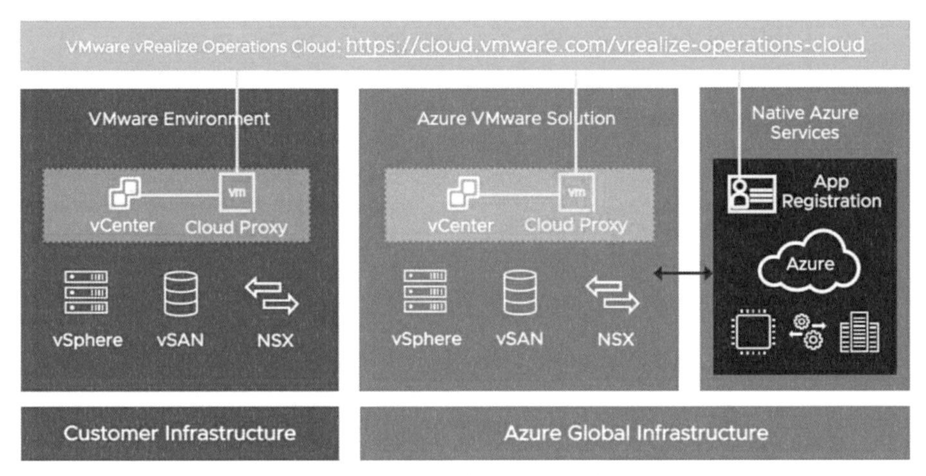

Fig. 1. vRealize Operations in multi-cloud management.

1.2 Notes on Methodology

In our study we apply both regression and classification methods, including rule induction algorithms, as well as information-theoretic feature ranking techniques. In one scenario, we are interested in identifying potential factors explaining the KPI behavior, in another one, the problem is to predict KPI abnormality and deduce "rules" leading to such situations. Rule induction is always preferable if the interpretability of the models and patterns are required compared to their predictive power. In this work we experimented with CN2 [6] (see its implementation in the visual programming tool Orange [7]).

Further research needs to address the problem of efficient management of multiple trade-offing KPIs for an application.

2 Acknowledgement

Authors are thankful to the reviewers for valuable feedback. Particularly, references [8] and [9] on RCA topic were recommended to our attention which treat special uses cases of cellular networks and cloud databases management, respectively, and are helpful in our further research on this subject.

Arnak Poghosyan was funded by RA Science Committee, in the frames of the research project № 20TTAT-AIa014.

References

1. Sole, M., Muntes-Mulero, V., Rana, A.I., and Estrada, G.: Survey on models and techniques for root-cause analysis. arXiv: 1701.08556v2, (2017).
2. Barredo Arrieta, A., Díaz-Rodríguez, N., Del Ser, J., Bennetot, A., Tabik, S., Barbado, A., Garcia, S., Gil-Lopez, S., Molina, D., Benjamins, R., Chatila, R., & Herrera, F.: Explainable Artificial Intelligence (XAI): Concepts, taxonomies, opportunities and challenges toward responsible AI, Information Fusion 58(6), 82-115 (2020).
3. Ribeira, M.T., Singh, S., Guestrin, C.: Why should I trust you?: Explaining the predictions of any classifier. arXiv: 1602.04938v1 (2016).
4. VMware vRealize Operations Manager: https://www.vmware.com/products/vrealize-operations.html.
5. VMware products, https://www.vmware.com/products.html.
6. Fürnkranz, J., Gamberger, D., Lavrac, N.: Foundations of rule learning. Springer-Verlag, Heidelberg (2012).
7. Orange Software: https://orangedatamining.com/.
8. Mdini, M: Anomaly Detection and root cause diagnosis in cellular networks. PhD Thesis. Rennes (2019).
9. Ma, M, Yin, Zh, Zhang, Sh., Wang, Sh., Zeng, Ch., Jiang, X., Hu, H., Luo, Ch.: Diagnosing root causes of intermittent slow queries in cloud databases. PVLDB, 13(8), 1176-1189, 2020.

Root Cause Analysis of Application Performance Degradations via Distributed Tracing

Arnak Poghosyan[1,2][0000-0002-6037-4851], Ashot Harutyunyan[1,3,4][0000-0003-2707-1039], Naira Grigoryan[1][0000-0003-3980-4500] and Clement Pang[1][0000-0002-5821-0735]

[1] VMware, Inc., Palo Alto, CA 94304, USA
{apoghosyan; aharutyunyan; ngrigoryan@vmware.com; clementp@gmail.com}
[2] Institute of Mathematics of NAS RA, 0019 Yerevan, Armenia
[3] Institute of Informatics and Automation Problems of NAS RA, 0014 Yerevan, Armenia
[4]AI Lab at Yerevan State University, 0025 Yerevan, Armenia

Abstract. Diagnostics of IT issues in cloud applications by means of the distributed tracing is one of the key functionalities of modern monitoring and management systems. Traces are one of the main data sources for AI/ML empowered analytics together with time series and logs. Proactive identification and accelerated remediation of issues in complex cloud environments require real time and intelligent root cause analysis based on all available data sources before those problems will affect end-users. However, successful adoption of AI solutions is anchored on trust and the acknowledgement of potential benefits. Explainable AI (XAI) is gaining more and more popularity by enabling improved reliability, confidence and trust towards smart solutions. We focus our attention to modern applications with microservices architecture and consider the application of rule-induction systems to the tracing traffic passing through a malfunctioning service for possible explanations. Rule-learning classification methods have sufficient scalability and interpretability for revealing practical patterns from large datasets. We show how those methods identify microservices responsible for the performance degradations.

Keywords: Explainable AI, rule learning algorithms, distributed tracing, root cause analysis.

1 Introduction

Cloud computing is the reality of modern IT infrastructures and applications. It provides with the plenty of benefits like scalability, resiliency, security and elasticity. However, it also has several complications which will alleviate the benefits if not appropriately handled. The first and foremost is the cloud architecture complexity with heterogeneous components which makes unrealistic for system administrators to perform manual optimizations and troubleshooting. Cloud systems require continuous monitoring with AI/ML empowered analytics for timely prevention of upcoming performance degradations.

The growth of large-scale distributed cloud environments requires more advanced and intelligent root cause analysis (RCA) of IT issues. System administrators are no longer able to perform real-time decision making and explainable RCA should be the most valuable property of AI engines. Explainable AI (XAI) [1] builds trust and increases the expertise for a better system administration. The main purpose of XAI in management solutions is not only in timely predictions of upcoming IT issues but primarily in explaining why and how those decisions were made. The list of powerful XAI solutions is rather long. It includes modern methods like LIME [1], SHAP [2], the classical rule-induction systems [3] and tree-based methods [5], [6].

Distributed tracing (see [7-9]) is the best-known method for the monitoring of modern applications especially consisting of microservices architecture. Traces (see Fig. 1) observe end-to-end requests that propagate through the distributed microservices and detect transaction slowdowns. A single trace contains a series of tagged time intervals known as spans. The spans contain some metadata (known as dimensions or tags) for better process resolution. The tags contain more detailed information regarding the micro-process and may include information about a user, the process duration, start time, cancel time, server IP, etc. A tracing traffic shows how applications and services interact. The monitoring of a tracing traffic through specific services can be accomplished via RED metrics (see Fig. 1). RED stands for: (R) request rate - showing the number of requests per second, (E) error rate - showing the number of erroneous traces per second, and (D) request durations. They identify the status of a system in terms of requests, errors and durations showing whether the system is slow or has a lot of errors. The main purpose of explainable RCA is detection of responsible micro-services (spans) and their tag-values when the corresponding RED metrics indicate problems in specific micro-services.

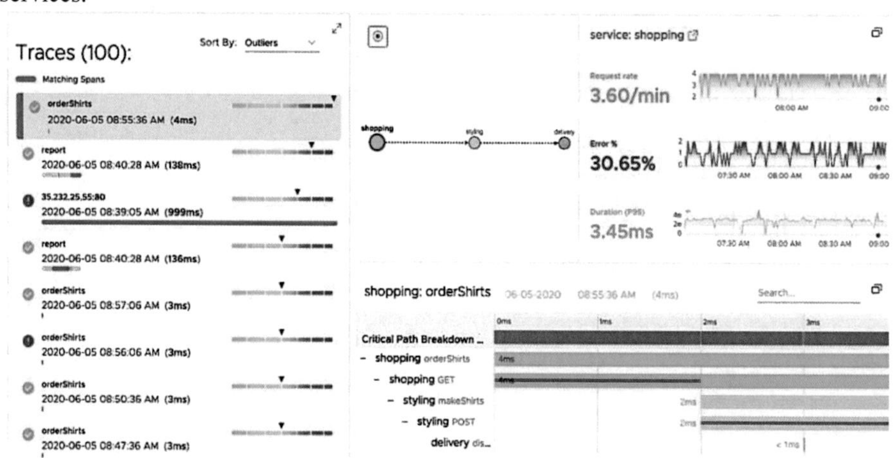

Fig. 1. Tracing browser for an application troubleshooting. RED metrics monitor the health of microservices through request rates, error rates and durations of traces.

The visualization of a tracing traffic is known as application map (see Fig. 2). We can explore specific services, view requests, errors, and durations for each service and see the tracing traffic, including the traffic directions. RED metrics for specific services identify the statuses in different colors. By default, the statuses are visualized based on the errors. The traces corresponding to the blue-colored services have a few errors. The traces corresponding to the red-colored services have a big portion of errors. By navigating into a degraded node (see the red-colored nodes in Fig. 2), we can collect all available traces passing through the selected service and acquire insights regarding the causes of the problems.

Analysis of a tracing traffic for troubleshooting purposes were discussed in literature for several applications (see [10-12]). We focus our attention to modern cloud applications with microservices architecture monitored via distributed tracing. The RCA can be triggered simultaneously based on information on errors, requests and durations. In the first case, the labeling of traces will be performed via erroneous traces. In the second case, the labeling will be performed via trace durations. Here, we need to identify the normality for a specific class of traces and label correspondingly the traces with smaller or bigger durations. In the third case, the labeling should be performed via request information.

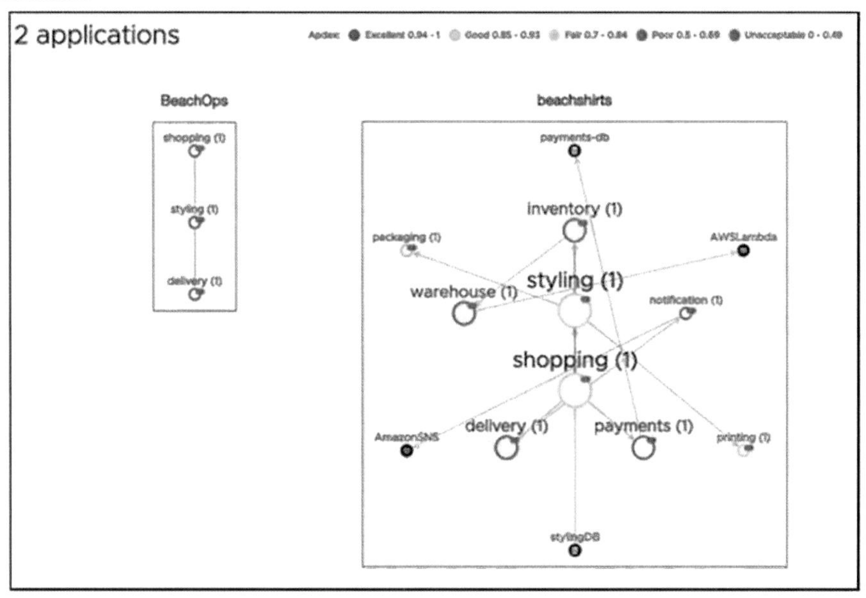

Fig. 2. An application map for a trace-traffic visualization. Colors indicate the health of each service based on the values of RED (requests, errors, and durations) metrics. By default, the services are colored based on errors.

We discuss the application of rule induction classification algorithms to a tracing traffic for identifying potential conditions that explain issues via trace types, spans, tags and tag-values. Rule learning (see [3]) is a form of XAI. Its power is in knowledge discovery and explainability. Rule learners are the best if simplicity and human understandability are superior compared to the predictive power.

RIPPER (see [13]) is still the state-of-the-art in inductive rule-learning. It supports missing values, numerical and categorical variables, multiple classes, and scales nearly linearly with the number of instances in a dataset. The goal of the RIPPER is finding regularities in data in the form of an IF-THEN rule. Those rules will contain detailed information on a specific microservices indicating the corresponding names and related tag-values. We explored several solutions. One of them identifies only the service names responsible for the performance degradations. This is the fastest solution as the corresponding tabular data is the smallest, and it is easy to process. However, it has lower level of explainability. The other solution processes both the names and the corresponding tag-values. It has higher level of explainability, but is slower as the corresponding tabular data contains span-names and their tag-values. The total number of tag-values can be rather big for different applications. Fig. 3 shows the result of application of the first approach. We hid the corresponding span-names due to confidentiality.

We performed experiments for one of VMware customers. We detected a spike in one of the RED metrics responsible for the errors for a specific service. The corresponding service in the application map turned into the red color. We queried around 11,000 traces passing through that service for the latest minutes when the spike was detected. Around 2000 traces were erroneous with label “output = 1” that should be explained. The traces had 186 unique span names and the goal was detection of those ones responsible for the errors. Fig. 3 shows five such rules revealed by RIPPER. The rules are simple containing only one condition.

(span A = 1) ⇒ output = 1 (868/17)
(span B = 1) ⇒ output = 1 (501/0)
(span C = 1) ⇒ output = 1 (511/211)
(span D = 1) ⇒ output = 1 (185/25)
(span E = 1) ⇒ output = 1 (127/3)

⇒ output = 0 (9701/84)

Fig. 3. An example of rules induced via RIPPER.

The fraction at the end of each rule shows the number of traces firing the rule and the number of misclassifications. For example, the first rule will be fired for 868 traces with 17 false positive alarms. The fractions allow us to calculate the recall and precision of each rule. Different rules can be prioritized via recall and/or precision.

2 Conclusion

We showed how rule-induction methods can be applied for the root cause analysis of an application performance degradations. We focused our attention to modern cloud applications with microservices architecture monitored via distributed tracing. Our goal was providing explanations showing which microservices were responsible for a transaction failure or slowdown and, when possible, revealed detailed information regarding the causes of a problem. We discussed the application of RIPPER to this problem and its potential benefits.

3 Acknowledgement

Arnak Poghosyan was funded by RA Science Committee, in the frames of the research project № 20TTAT-AIa014. We thank anonymous reviewers for careful and critical reading of our paper.

References

1. Arrieta, A.B., Díaz-Rodríguez, N., Ser, J.D., Bennetot, A., Tabik, S., Barbado, A., Garcia, S., Gil-Lopez, S., Molina, D., Benjamins, R., Chatila, R., Herrera, F.: Explainable artificial intelligence (XAI): Concepts, taxonomies, opportunities and challenges toward responsible AI. Information Fusion, vol. 58, pp. 82-115 (2020).
2. Ribeiro, M. T., Singh, S., Guestrin, C.: "Why should I trust you?": Explaining the predictions of any classifier. arXiv: 1602.04938v3. https://arxiv.org/abs/1602.04938, 2016.
3. Lundberg, S., Lee, S.-I.: A unified approach to interpreting model predictions. arXiv: 1705.07874. https://arxiv.org/abs/1705.07874, 2017.
4. Fürnkranz, J., Gamberger, D., Lavrac, N.: Foundations of rule learning. Cognitive Technologies, Springer, Heidelberg, 2012.
5. Louppe, G.: Understanding Random Forests: From theory to practice, PhD Thesis, U. of Liege, 2014.
6. Breiman, L.: Random Forests, Machine Learning, vol. 45, N 1, pp. 5-32 (2001).
7. VMware Tanzu Observability: Distributed tracing overview. https://docs.wavefront.com/tracing_basics.html.
8. Distributed tracing: A complete guide. https://lightstep.com/distributed-tracing.
9. Zipkin distributed tracing system. https://zipkin.io.
10. Ezzati-Jivan, N., Daoud, H., Dagenais, M.R.: Debugging of performance degradation in distributed requests handling using multilevel trace analysis. Wireless Communications and Mobile Computing, vol. 2021, Article ID 8478076, 17 pages (2021).
11. Cassé, C., Berthou, P., Owezarski, P., Josset, S.: Using distributed tracing to identify inefficient resources composition in cloud applications. IEEE 10th International Conference on Cloud Networking (CloudNet2021), Nov 2021, Cookeville, TN, United States.
12. Cai, Z., Li, W., Zhu, W., Liu, L., Yang, B.: A real-time trace-level root-cause diagnosis system in Alibaba datacenters. IEEE Access, vol. 7, pp. 142692-142702 (2019).
13. Cohen, W.W.: Fast effective rule induction. In: Proceedings of the Twelfth International Conference on Machine Learning, pp. 115–123 (1995).

On AI-Driven Customer Support in Cloud Operations

Ashot Baghdasaryan[1,2][0000-0003-1424-0382], Tigran Bunarjyan[1][0000-0003-4427-4284], Arnak Poghosyan[1,4][0000-0002-6037-4851], Ashot Harutyunyan[1,2,3][0000-0003-2707-1039] and Jad El-Zein[1]

[1] VMware, Inc., Palo Alto, CA 94304, USA
[2] AI Lab at Yerevan State University, 0025 Yerevan, Armenia
[3] Institute for Informatics and Automation Problems of NAS RA, 0014 Yerevan, Armenia [4] Institute of Mathematics of NAS RA, 0019 Yerevan, Armenia

{ashotb; tbunarjyan; apoghosyan; aharutyunyan; jelzein@vmware.com}

Abstract. Proactive and timely user support is the foundation of customer loyalty. One of the main reasons of customer loyalty and satisfaction is the tolerable mean time to resolution of service requests. Another important measure of successful customer support is the reduction of the total number of service requests. Those key characteristics can't be improved via extensive time investments by technical support engineers due to the complexity of modern cloud operations. There is a natural lack of domain expertise due to the diversity and complexity of customer ecosystems. The only feasible solution to the problem is the development of proactive and intelligent analytics that not only automate/accelerate the resolution of currently available requests but also anticipate potential issues through proper customer segmentation and rule discovery. We describe an approach to the problem via relevant recommendations which link together historical requests and the corresponding knowledge base article as a potential remediation of the identical issues. Moreover, it reveals trending issues and requests which require urgent intervention of technical engineers. Based on the provided recommendations, they can construct rules for the proactive resolution of similar issues in a customer environment from the same segment. The AI-engine leverages several pipelines for data collection and preprocessing, model training and knowledge discovery. Data pipelines process a large collection of service requests and knowledge base articles. Training pipelines learn product-specific language models based on BERT transformers which the recommender system utilizes for a problem resolution.

Keywords: Customer Support Service, Service Request, Knowledge Base Article, Proactive Support, NLP, BERT, Transfer Learning, Recommender System.

1 Introduction

A successful customer service is the future of any IT company as it supports the development of a business, the improvement of a retention and the growth of sales. Customer services follow two parallel strategies – the improvement of ongoing case-resolution processes by automation and acceleration of decision-making flows, and anticipation of the future trendy issues with proactive resolutions. Both objectives can be accomplished via AI-driven analytics based on the behavioral analysis of customers, cross-customer data correlations, historical analysis of previous service requests (SR) and their resolution experience via knowledge base (KB) articles, expert knowledge for a "rule" generation, etc. Cloud operations are so complex, and the number of daily issues is so big that the simple increase of the number of support engineers will be insufficient for the reduction of the mean time to resolution (MTTR). The development of AI platforms is the necessity for the modern enterprise cloud applications where the expert knowledge of support engineers can be an important complement.

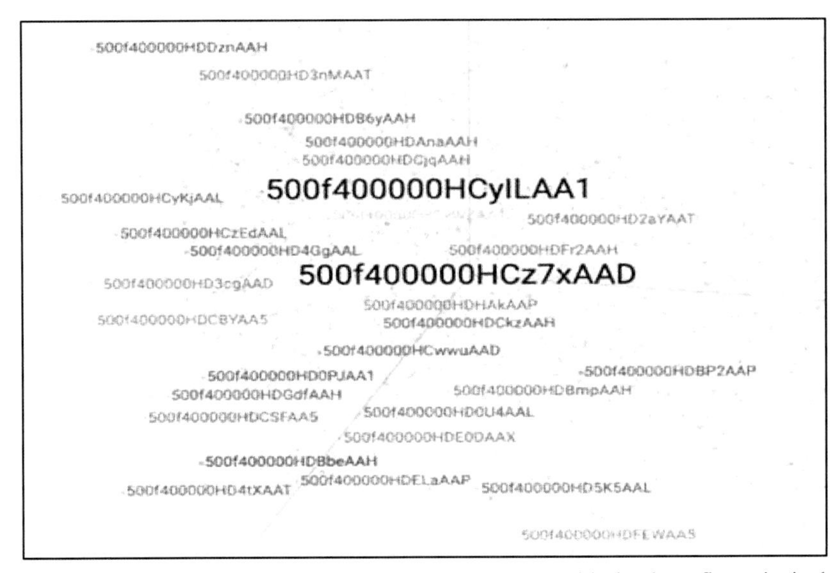

Fig. 1. Visualization of feature vectors in the 3D space with the three first principal components.

Our main goal is to demonstrate the process of mitigation of the complexity of support operations and the improvement of the proactive support capabilities with self-remediation perspectives. The solution employs a series of pipelines with the final goal to train a product-specific language model based on historical SR and KB datasets. Language models will assign to each SR and KB some feature vectors (see Fig. 1) for similarity comparisons. We utilize cosine similarity distance and similar objects must have almost collinear feature vectors with cosine distance equal to 1. Fig. 1 shows such two similar SR documents with the highlighted IDs which are also close in the 3D space of the first three principal components.

The corresponding recommender system can utilize the feature vectors for matching similar SRs and KBs (see direct KB recommendation box of Fig. 2) that will accelerate the process of resolution. It also can recommend similar SRs (see SR recommendation box of Fig. 2) that can be helpful for clarifying the nature of an unknown current problem by matching it with the known historical ones. The set of similar historical SRs can contain some historical KBs potentially useful for the current issue (see indirect KB recommendation box of Fig. 2). Meanwhile, the detection of recent similar issues from a specific customer group can help for identifying upcoming issues and proactively announce remediation strategies through the set of product and customer specific rules.

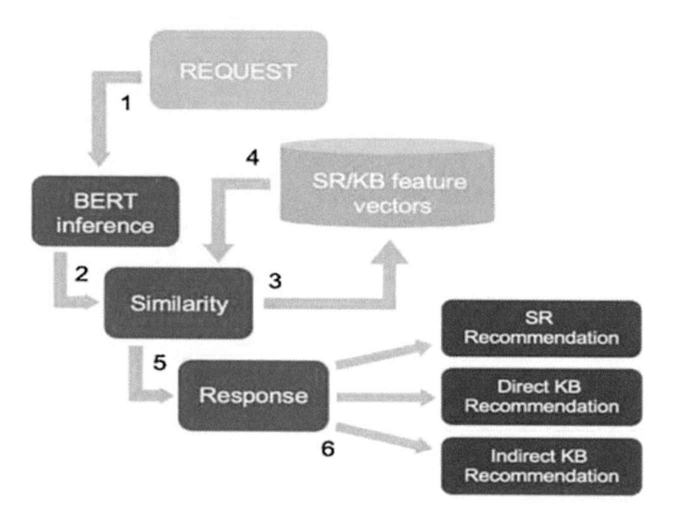

Fig 2. The high-level schema of the recommender system.

The recommender system utilizes Bidirectional Encoder Representations from Transformers (BERT) language model (see [1]). We are using BERT-Large pretrained model with 24-layers, 1024-hidden layers, 16-heads and 340 million parameters. BERT has "cased" and "uncased" models. We use uncased models as our datasets are insensitive towards the case information. There are two different techniques for word masking processes. The first one is the random selection of masks, and the second one is the whole word masking. In the latest models, all tokens corresponding to a word are masked at once. Experiments showed that the whole word masking technique slightly outperforms. Fine tuning of the pretrained BERT model is performed on SR and KB customer datasets containing more than a million documents. We perform tuning of the model with the sequence length of 128 for 90% of the steps. Then, we train the rest 10% with the sequence length of 512 (the final length of the feature vectors) to learn the positional embeddings. The experiments were showed that masked language model accuracy for fine-tuned model was about 77% while next sentence prediction accuracy was almost 95%.

2 Acknowledgement

Arnak Poghosyan and Tigran Bunarjyan were funded by RA Science Committee, in the frames of the research project № 20TTAT-AIa014.

References

1. Devlin, J., Chang, M.-W., Lee, K., Toutanova, K.: BERT: Pre-training of Deep Bidirectional Transformers for Language Understanding. In: Proceedings of the 2019 Conference of the North American Chapter of the Association for Computational Linguistics: Human Language Technologies, vol. 1 (Long and Short Papers), pp. 4171–4186. Association for Computational Linguistics, Minneapolis, Minnesota (2019).

Computer Vision Applications for Smart Cities Using Remote Sensing Data: Review

Lilit Yolyan[1] [0000−0002−7848−3124]

Yerevan State University, Alek Manukyan 1, 0025, Yerevan, Armenia
l.yolyan@ysu.am

Abstract. In a rapidly growing level of urbanization, overpopulation has become one of the main problems for municipalities, the country's economy, and public administration. Many issues related to waste management, urban resource planning, air pollution, traffic and traffic congestion, and public health issues challenge existing infrastructure. A smart city aims to improve the way people live, create a more sustainable environment, and make it easier for municipalities and governments to manage all of these processes. The application of artificial intelligence and computer vision methods can solve the problems of a smart city (surveillance, area coverage, land use and land cover, damage monitoring, fire detection, etc.) that were impossible or hardly solved a few years ago. Here we review deep learning and computer vision applications for smart cities using remote sensing data. In addition, we present 2 types of data sources for creating smart city datasets for cities in Armenia. The first type is potential data sources that can be collected through the efforts of the municipality, the second type is open source data ready to be used for the solutions below.

Keywords: Smart City · Computer Vision · Deep Learning · Remote Sensing · Artificial Intelligence.

1 Introduction

According to the World Bank dashboard, the world population in 2021 is 7.8 billion. At the same time, the level of urbanization in 2020 amounted to 56.15% [1]. Based on the UN Economic and Social Affairs report in 2050 the level will reach 68% [2]. Due to the overpopulation of existing cities, municipalities are not prepared to deal with the many different challenges that a high level of urbanization brings with it. Many issues related to waste management, urban resource planning, air pollution, traffic, transportation overload, and public health issues challenge existing infrastructure [3]. Cities are currently facing these challenges and are trying to find new, revolutionary and smart ways to deal with emerging and existing overpopulation problems. Although the concept of a smart city is becoming more and more trendy among researchers, there is no single clear definition of what this paradigm represents. Paskaleva K.A. (2009) defines a smart city as one that takes advantage of the opportunities offered by Information and Communication Technologies (ICT) in increasing local prosperity and competitiveness – an approach that implies integrated urban development involving multi-actor, multi-sector and multi-level perspectives [4]. A city "connects the physical infrastructure, the IT infrastructure, the social infrastructure, and the business infrastructure to leverage the collective intelligence of the city" [5]. A city "combining ICT and Web 2.0 technology with other organizational, design and planning efforts to dematerialize and speed up bureaucratic processes and help to identify new, innovative solutions to city management complexity, to improve sustainability and livability" [6].

In 2012, Chourabi introduced a completely new framework to explain smart city concepts, and based on this framework, there are 8 critical factors that can help create new smart city initiatives: management and organization, technology, governance, policy context, people and communities, economy, built infrastructure, and natural environment [3].

As we can see above, in different publications the concepts of a smart city are described in different ways, but one thing unites them. Most researchers note the importance of ICT in the development and implementation of smart city concepts in urban infrastructure. In Chourabi's framework technology is one of the 8 factors that can influence the development of the city.

As was mentioned before ICT is one of the game-changer in city planning, it has a huge impact on the development and implementation of smart city concepts. Technologies like sensors, IoT, networks, and algorithms can make cities smarter[7]. Those technologies can have a huge impact on all different infrastructures of the city like city administration, education, healthcare, public safety, real estate, transportation, and utilities. Harrison et al. mention the importance of operational data, which is a combination of data extracted from traffic, power consumption, and other sources, in order to optimize the operations. In their work three points are mentioned as the most important features of smart cities: (1) the near-real-time data obtained from physical and virtual sensors, (2) the interconnection between different services and technologies inside the city, and (3) the intelligence from the analysis of the data and the process of optimizing and visualizing it [5]. Sensing technologies and IoTs are the source of the huge volume of data that can be gathered and processed to solve problems like water distribution, electricity distribution, energy consumption in buildings, building and bridges monitoring, environmental monitoring, etc [22]. Some researchers stress the importance of data gathering, as a starting point for creating a smart city. Then this data can be used to provide services [8]. Gharaibeh and al. discuss different applications of smart cities using data gathered from sensor networks, Unmanned Aerial Vehicles (UAVs), Vehicular Ad hoc Networks (VANETs), IoT, Social Networks, 5G and Device-to-Device (D2D) communications [23].

2 Visual Data for Smart City Problems

The many definitions presented above show that ICT is one of the keys to success in making a city smart. But collecting data is only a small part of the whole process. Gathered data can be used differently: (1) visualization mechanisms like dashboards for different e-government departments [9], (2) video streaming from surveillance cameras for safety reasons [10,11], (3) monitoring systems for buildings and roads [12], (4) sensors for transportation improvements [13].

A huge part of the data for the smart city can be visual data gathered from (1) surveillance cameras, (2) city cameras, (3) Unmanned Aerial Vehicles (UAV), (4) Satellites, etc. Visual data can be used for data visualization and AI in different areas.

One of the most useful data sources can be aerial imageries, that gather remote sensing data. "Remote sensing is the practice of deriving information about the Earth's land and water surfaces using images acquired from an overhead perspective, using electromagnetic radiation in one or more regions of the electromagnetic spectrum reflected or emitted from the Earth's surface" [14].

Remote sensing data is collected with the help of special cameras which measure reflected and emitted radiation of an area at a distance. They use different technologies like photogrammetry or LiDAR sensors to find fingerprints of objects on the Earth surface. Based on the type and the used technology these cameras can be placed on:

- Satellites and airplanes and can collect images of very large areas on the Earth surface,
- Sonar systems on the ships can scan the ocean floor without reaching the bottom,
- Cameras on satellites can be used to capture temperature changes in oceans.

Types of remote sensing data differ depending on camera type and where it will be placed. Those are different types of remote sensing data:

- Light Detection and Ranging (LIDAR),
- Radio Detection and Ranging (RADAR),
- Unmanned Aerial Systems (UAS),
- Hyperspectral Imagery,
- Thermal Imagery,
- Aerial Photography.

LIDAR cameras use laser signals to capture objects on the surface of the earth, RADAR systems work with radio waves instead of a laser. The third type uses Unmanned Aerial Vehicles or other similar technologies to capture images from the bird's height. Hyperspectral Imageries analyze a wide spectrum of light instead of capturing images in RGB. Thermal systems use heat to identify and visualize objects. Aerial images are very similar to UAS images, the main difference is that aircraft, rockets, or other spacecraft are used to capture images.

3 Machine Learning and Computer Vision Applications in Smart Cities

Remote sensing data is collected in huge quantities. The correct use of such a volume of remote sensing data can become a solution to various smart city problems. And some of these solutions can be found using AI and machine learning methods. Machine learning algorithms (Support Vector Machines (SVM), Ensemble models) were used in early applications of AI in remote sensing. Nowadays the research involves more and more applications of deep learning. First damage detection, classification, and localization problems are solved with algorithms like SVM, and later Convolutional Neural Networks (CNN) as well [15]. A huge amount of smart sensors and Internet of things (IoT) data was gathered in the sector of energy consumption and mobility where classical machine learning algorithms are more useful.

Cugurullo suggests three categories of urban artificial intelligence. The first one is associated with autonomous cars, which can result in many urban changes like decreasing traffic, energy use, etc. The second category is robotics, which can improve and fasten many mechanical processes. And last but not least is the city brain which is a monitoring mechanism that can collect data from different IoT sensors, analyze the data and create more data-driven decision-making processes [16].

Deep learning techniques are one of the most popular ones when it comes to AI in smart cities. The subfield of it is convolutional neural networks which can be very effective in solving different computer vision techniques. CNNs are the state of the art for image data. Aerial, remote sensing imageries and surveillance videos can be input data for computer vision models.

Classification and object detection networks can be used for rescue monitoring, destruction monitoring, area coverage calculations, environment mapping, and surveillance. Drone data can be used in smart parking, fire detection, environmental clean-up, agricultural monitoring, etc [17]. Few-shot learning techniques [20] and Siamese networks [21] can be used for image registration.

City cameras can be used in authentication in different urban areas (subways, airports, and other public places. For security reasons, car license plate identification and face recognition techniques can be applied [18]. Calculating traffic volume and driving quality estimation can be done using city cameras or UAV data. City cameras are a reliable source of data, however, they may have issues with coverage. UAVs can be used to collect data from regions where it's difficult to deploy other technologies [23]. Siamese networks can be used in image registration problems [19]. Table 1 shows the types of smart city solutions and the computer vision techniques needed to implement them.

4 Implementation for Yerevan

As was mentioned above there are 4 main remote sensing data sources that can be used for implementing computer vision techniques to solve smart city problems: UAVs, city cameras, aerial images, and satellite images. To collect UAV and aerial images some governmental funds and initiatives are needed. That's why for the rest of the paper we will discuss only accessible and existing data sources. Existing data sources for Yerevan are city cameras and satellite images. According to a 26.08.2020 report of the Police of RA states, there are 150 crossroad cameras in Yerevan. Many other cameras are placed from Yerevan Mu-

Table 1. Summary of computer vision application with remote sensing data for smart cities.

Problem Type	Computer Vision Technique	Remote Sensing Data Type
Rescue Monitoring	Object detection, segmentation	UAV, aerial images
Area Coverage, Land Usage, and Lang Coverage (LULC)	Object detection, segmentation	UAV, aerial images satellite images
Damage Monitoring	Classification, object detection	UAV, city cameras
Surveillance	Classification, object detection, anomaly detection	City cameras, UAV
Smart Parking	Classification, object detection	City cameras, UAV
Fire Detection	Classification	City cameras, UAV
Agricultural Monitoring	Object detection, Segmentation	UAV, aerial images, satellite images
People Authentication	Object detection, Siamese networks, Few shot learning	City cameras
License Plate Detection	Object detection	City cameras, UAV
Drivers Behavior Estimation	Anomaly detection	City cameras
Image Registration	Siamese network, Few-shot learning	UAV, aerial images, satellite images

nicipality showing aerial views of the Republic Square, Sasuntsi David Square, Cascade complex, North Avenue, Mashtots park, Charles Aznavour, and Garegin Nzhdeh Squares in Yerevan city. The main problem with these cameras is that data is lifestream and it's not collected for a longer timestamp. Moreover, the companies that own cameras don't have warehouses to keep that much data. The next data source mentioned above is satellite images. 3 satellites collect data from Armenia as well: Landsat-7, Landsat-8, and Sanitel-2. USA launched Landsat-7 on April 15, 1999. At the time it was the most accurately calibrated Earth-observing satellite. Landsat-8 is an American Earth observation satellite launched on 11 February 2013. Those two are the seventh and eighth satellites in the Landsat program. Landsat-7 and Landsat-8 have spatial resolutions of 30 meters. Sentinel-2 is an Earth observation mission from the Copernicus Programme that systematically acquires optical imagery at a high spatial resolution over land and coastal waters. The spatial resolution of Sanitel-2 is from 10 to 60 meters.

Table 2. Satellite Scene of Yerevan

Name	Number of scenes of Yerevan from 09/30/2020 - 09/30/2021	Monthly number of scenes of Yerevan	Coverage in km per scene
Landsat-7	44	3-4	185x180
Landsat-8	35	3-4	183x183
Sanitel-2	132	8-12	290x290

Table 2 represents the number of scenes(images) of Yerevan per satellite mentioned above. Here it is noticeable that satellites cover a huge area of land which are good for several types of problems described above (Area Coverage, LULC, Agricultural monitoring, etc), but may not be possible to use in other cases (Damage Monitoring, Surveillance, Fire Detection, etc.)

5 Conclusion

A huge part of smart city problems can be defined as machine learning and deep learning problems, but main differentiations are underlying in data type, size, and structure. Here AI problems are more unique, as domain knowledge and data specialties create new difficulties that are different from classic, general-purpose AI problems.

Smart City solutions are new for Yerevan. Concepts and potential implementations are discussed at many datatons and other conferences. However, real implementations are lacking. Only small projects were developed, such as smart parking lots with camera detection, educational platforms. Many small projects and groups are working on specific solutions such as smart garbage collection, smart traffic, etc. However, the lack of procedures and well-designed plans supported by the government hinders the implementation of more general solutions.

However, Yerevan has great potential to use the large amount of data already collected to solve the problems presented in the previous section.

Smart city challenges cannot be solved without proper data, and remote sensing is one such data source that can be most useful. Remote sensing data enables several smart city challenges to be addressed, ranging from city monitoring with dashboards to AI applications for more complex tasks. However, collecting and working with such big data can be a challenge. The examples in Yerevan showed that good infrastructure is needed to create processes for working with such data. Funding from the government or municipality is important to establish proper data collection procedures and warehouses to collect data from UAVs, aerial images, and city cameras. In addition, it is important to create well-defined procedures and problems to solve with already available data (eg open satellite imagery).

However, existing datasets can help with AI tasks such as area coverage and land use prediction, agricultural monitoring, water quality monitoring, etc. They can be the first steps towards more technological and smart cities without requiring huge resources.

References

1. The World Bank Website, https://data.worldbank.org/indicator/SP.URB.TOTL.IN.ZS last accessed 2022/05/19
2. World Urbanization Prospects The 2018 Revision, https://population.un.org/wup/Publications/Files/WUP2018-Report.pdf, last accessed 2022/05/19
3. Chourabi, H., Nam, T., Walker, S., Gil-Garcia, J.R., Mellouli, S., Nahon, K., Pardo, T.A., Scholl, H.J.: Understanding smart cities: An integrative framework. 2012 45th Hawaii International Conference on System Sciences (2012)
4. Paskaleva, K.: Enabling the smart city: the progress of city e-governance in Europe. International Journal of Innovation and Regional Development (2009)
5. Harrison, C., Eckman, B., Hamilton, R., Hartswick, P., Kalagnanam, J., Paraszczak, J., Williams, P.: Foundations for Smarter Cities. IBM Journal of Research and Development (2010)
6. Toppeta, D.: The smart city vision: how innovation and ICT can build smart, "livable", sustainable cities. The Innovation Knowledge Foundation 5, pp 1-9 (2010)
7. Sánchez-Corcuera, R., Nuñez-Marcos, A., Sesma-Solance, J.: Smart cities survey: Technologies, application domains and challenges for the cities of the future. International Journal of Distributed Sensor Networks (2009) doi:10.1177/1550147719853984
8. Al-Hader, M., Rodzi, A., Sharif, A. R., Ahmad, N: Smart city components architecture. 2009 International Conference on Computational Intelligence, Modeling and Simulation, pp. 93-97 (2009)
9. Jing, C., Du, M., Li, S., Liu, S.: Geospatial Dashboards for Monitoring Smart City Performance. Sustainability (2019). https://doi.org/10.3390/su11205648
10. Nigam N., Kroo I.: Persistent Surveillance Using Multiple Unmanned Air Vehicles. The Institute of Electrical and Electronics Engineers Aerospace Conference (2008)

11. Tuyishimire, E., Adiel, I., Rekhis, S., Bagula, B. A., Boudriga, N.: Internet of things in motion: A cooperative data muling model under revisit constraints. 2016 International IEEE Conferences on Ubiquitous Intelligence and Computing, Advanced and Trusted Computing, Scalable Computing and Communications, Cloud and Big Data Computing, Internet of People, and Smart World Congress, pp. 1123-1130 (2016)
12. Li, D., Shan, J., Shao, Z., Zhou, X., Yao, Y.: Geomatics for smart cities-concept, key techniques, and applications. Geo-spatial Information Science, pp. 13-24 (2013)
13. Nikitas, A., Michalakopoulou, K., Njoya, E. T., Karampatzakis, D.: Artificial intelligence, transport and the smart city: Definitions and dimensions of a new mobility era. Sustainability (2020)
14. Campbell, J. B., Wynne, R. H.: Introduction to remote sensing. Guilford Press (2011)
15. Luckey, D., Fritz, H., Legatiuk, D., Dragos, K., Smarsly, K.: Artificial intelligence techniques for smart city applications. International Conference on Computing in Civil and Building Engineering, pp. 3-15. Springer, Cham (2020)
16. Cugurullo F.: Urban Artificial Intelligence: From Automation to Autonomy in the Smart City. Frontiers in Sustainable Cities, (2020)
17. Ismail, A., Bagula, B. A.,Tuyishimire, E.: Internet-of-things in motion: A UAV coalition model for remote sensing in smart cities. Sensors (2018)
18. Shervan E.F.: Computer Vision in Smart City. International Festival on Automation and Computer, Isfahan, Iran (2019)
19. Ma, L., Liu, Y., Zhang, X., Ye, Y., Yin, G., Johnson, B. A.: Deep learning in remote sensing applications: A meta-analysis and review. ISPRS Journal of Photogrammetry and Remote Sensing (2019)
20. Wang, Y., Yao, Q., Kwok, J. T., Ni, L. M.: Generalizing from a few examples: A survey on few-shot learning. Association for Computing Machinery Computing Surveys (2020)
21. Koch, G.R.: Siamese Neural Networks for One-Shot Image Recognition (2015)
22. Hancke, G. P., de Carvalho e Silva, B., Hancke G. P.: The role of advanced sensing in smart cities. Sensors (2012)
23. Gharaibeh, A., Salahuddin, M. A., Hussini, S. J., Khreishah, A., Khalil, I., Guizani, M., Al-Fuqaha, A.: Smart cities: A survey on data management, security, and enabling technologies. IEEE Communications Surveys and Tutorials (2017)

Exploring the Feasibility of Video Activity Reporting for Students in Distance Learning

Zhen He[1], Sayan Sarcar and Tomoo Inoue[1][0000-0003-3600-214X]

[1] University of Tsukuba: Tsukuba, Ibaraki, JP
inoue@slis.tsukuba.ac.jp

Abstract. Activity reporting becomes important in remote work. It is an approach generally practiced in organizations for getting work progress. Video activity reporting has been proposed for distributed organizations, instead of conventional text activity reporting. Rich interactions realized by video activity reporting is expected to improve employees' work engagement, while it is informative to provide work engagement indicators.
In this article, we report the impressions on video activity reporting by university students. We conducted four or more weeks of trial with eight participants, and collected 232 video reports with engagement surveys. Subsequently, we interviewed six participants to explore the feasibility of video activity reporting. This work highlights the opportunities that video activity reporting can bring in remote work.

Keywords: Remote Work, Activity Report, Short Video, Work Engagement.

1 Introduction

Today, employees are no longer tied to their cooperate offices. Remote work has become commonplace. Compared with face-to-face working, flexible working hours and free workplaces in remote work bring benefits such as reduced work pressure and increased productivity [1]. However, with the increased acceptance of remote work, some issues are coming to the fore. Specifically, the decline in work engagement in remote work has become even more acute than face-to-face work [2, 3].

Work engagement, or often known as employee engagement, is described as a positive, fulfilling, and work-related state of mind [4]. Engaged employees, on the one hand, can support their organizations positively, resulting in higher productivity [5], better employee performance [6], and business success [7]; and on the other hand, positive work progress influences employees' own inner work-life significantly. Thus, maintenance of work engagement is always an essential topic for organizational management aspect.

In industry, a well-known and generally practiced approach for organizational management is activity reporting [8]. Employees regularly record their work progress and report it to their managers, and managers recognize employees' work progress through it.

During the process, employees also have a chance to review their old progress. In fact, not only employees, but it is also used by students in academics [9, 10]. In a word, activity reporting is widely applied through different time frames and for various purposes.

Activity reporting also brings various positive effects. Recent research found that recording daily work progress highlights can improve workplace wellbeing and inner work life, including work engagement [11]. Particularly under the remote work situation where engagement tends to decrease, activity reporting becomes even more essential.

On the other hand, existing activity reporting approaches are mostly text-based [12]. Compared with text, recently video has gain people's eyes more, which includes more information. In remote work, where communication is lack, the informative video can bring more opportunities to the distributed organizations. Followed by this idea, the use of video activity reporting has been proposed [13]. It suggests remote workers record videos instead of text to report their work progress, and their managers give feedback after check. Through the interaction based on video activity reporting, the internal communication in distributed organizations can be increased, and remote workers' engagement is expected to improve.

The preliminary findings in [13] also demonstrated the relationship between video activity reporting and engagement. Based on long-term daily video reports from one employee, the analysis proved that non-verbal features in video activity reporting have the potential to indicate one's engagement level. However, a few questions remain unanswered: (1) how such non-verbal features in video activity reporting provide a deeper understanding of engagement, and (2) how others than employees use video activity reporting.

Towards addressing these questions, we conduct a video activity reporting trial with eight university students in this work. After at least four weeks of continuous daily recording trial, we interview six participants to investigate the feasibility of video activity reporting for students. Meanwhile, we analyze a total of 232 daily video reports collected in the trial to generalize and extend the knowledge about non-verbal features that can become engagement indicators.

The qualitative analysis based on semi-structured interviews explores the benefits and limitations of video activity reports for students, while investigates the relationship between engagement and characteristic behaviors in video activity reporting.

2 Related Work

2.1 Work Engagement

Schaufeli et al. described work engagement as a positive, fulfilling, and work-related state of mind characterized by vigor, dedication, and absorption [4]. Although many other definitions of work engagement exist, all of them point out the importance of it. It can bring benefits to both organizations and individuals from various aspects [5, 6, 7]. On the other hand, work engagement can also refer to the academic students perform. The construct of work engagement proposed by Schaufeli et al. in [4] is applicable to students [15]. Since the activities students perform are similar to those of employees, which are goal-directed and structured, they can also be considered a kind of "work" in psychology [16]. Similarly, higher work engagement has also positively affected students, resulting in improved involvement in studies [17], and better academic performance [18].

Since work engagement can bring various positive effects, maintaining and promoting it is always critical for any organization. One simple but effective way is to sense and address disengagement timely. According to the study of how work engagement spread in a large organizational network, Mitra et al. found engagement and disengagement can spread from one employee to another [19]. Therefore, to maintain overall engagement, organizations are required to understand individuals' engagement declined issues promptly. However, in remote work that members cannot meet face-to-face, recognizing and measuring one's engagement becomes a problem.

2.2 Work Engagement Measurement

Work engagement is traditionally measured through surveys. In practice, many kinds of scales are used for engagement measurement. For example, the Utrecht Work Engagement Scale (UWES) measures engagement from three aspects of vigor, dedication, and absorption [4]. Intellectual, Social, Affective Engagement Scale (ISA) is based on the view that engagement has three dimensions: intellectual, social, and affective [20]. Gallup's Q12 employee engagement scale (GWA) uses 12 items to measure processes and issues that are actionable at the work group's supervisor or manager [21]. Job Engagement Scale (JES) followed the three components of engagement: physical, cognitive, and affective [22].

While various surveys have been demonstrated their effectiveness in measuring engagement, one of the common problems is timeliness. The periodic surveys (e.g., annual) generally practiced in organizations caused the difficulty in comprehending real-time engagement [23]. Since disengagement can spread from one to others [19], more real-time engagement measures are required in organizational management. Recent studies indicated social media text has the power to predict one's engagement [23]. Analyzing the word choice used in social media can provide more real-time insights into work engagement, allowing organizations to recognize and address engagement issues faster. Nevertheless, pulse waves, eye movements, and movements collected by wearable devices were also proved as potential real-time engagement indicators [24]. While these findings bring new opportunities to organizations, whether such measurements are suitable for remote work still needs discussion.

2.3 Activity Reporting

Activity reporting in the workplace is considered an effective organization management tool and essential communication among employee-manager pairs [8]. It is commonly used as formal communication of work progress, performing in various time frames (e.g., daily, weekly, and monthly) by modern organizations [12]. Recent works put the view on using technologies to support activity reporting. For example, a personal online tool was designed to help employees' summarize their daily highlights in work [11], and a conversational agent was proposed to promote employees' activity recording [25]. These tools bring the activity reporting positive impacts on engagement and self-learning. Besides, activating reporting itself is also powerful. The reporting content can be used to predict who will leave an organization [26] and provide engagement indicators [11].

Nevertheless, activity reporting is also used in academics by students. Activity reporting allows supervisors to receive information from students, which helps supervisors understand students' difficulties promptly and alter their instruction based on students' requirements. Activity reporting enables students to reflect on their understanding of the material and earn better learning skills [9]. Ito et al. discussed the utilization of weekly activity reporting for university students on system development PBL (Project-Based Learning) [10]. Their findings also demonstrated that word choice in PBL reports can be used to evaluate students' learning.

Although activity reporting is widely used, it is practiced chiefly in written and text-based [12], while media use other than text is limited. Therefore, the use of video activity reporting in remote work has been proposed [13]. By analyzing long-term video reports from an employee, the preliminary result suggests that non-verbal features could indicate one's engagement, bringing new opportunities to measure real-time engagement to distributed organizations. Also, the feedback from managers is proposed to promote remote workers' engagement, which aims to address the engagement declined issue in remote work.

Due to the similarity of students and employees in activity reporting and work engagement, how students practice video activity reporting is worth exploring. Through this work, we explored the feasibility of video activity reporting for students.

3 Video Activity Reporting Trial

In this work, we first conducted a video activity reporting trial with eight university students. After at least four weeks of continuously daily recording, 232 video reports were collected, as well as daily engagement surveys.

3.1 Participants

We enrolled eight graduated students from the University of Tsukuba through SNS. Due to the Covid-19 situation, all participants took lectures and did research remotely, which can be treated as remote work. As shown in Table 1, all participants are male, and ages were distributed from 23 to 26. For this trial, seven participants were international students from China, and one was native Japanese. Since Japanese language skills are required for reporting, non-native Japanese speaker participants all have at least JLPT N2 level Japanese proficiency [27], which can support them reporting in Japanese with no barriers. For each data provided, including daily video recording and engagement survey, we paid 100 Japanese yen to participants as rewards.

Table 1. Demographic statistics for participants and their trial results

Participant	Nationality	Gender	Age	Actual Trial Days/Video Reports
P1	Chinese	Male	23	63/56
P2	Chinese		25	24/23
P3	Chinese		25	36/33
P4	Chinese		25	34/33
P5	Chinese		23	37/37
P6	Chinese		24	16/16
P7	Chinese		26	16/16
P8	Japanese		23	26/18

3.2 Procedure

Participants were required to record a video every weekday (except national holidays in Japan), reporting their daily progress in Japanese. In detail, we instructed as follows: (1). Reporting content should be your study matters or your research progress. For example, the lecture you take, the meeting or seminar you attend, your current research progress and plan, the difficulty you meet, and the guidance from your supervisor. (2). For each report, 30 seconds around length is appropriate, but longer or shorter is also acceptable. (3). There is no limitation to the recording device. You can use your device (e.g., smartphone, computer) for recording. However, you need to take off accessories (e.g., masks) that can shade your face, ensuring your whole face can be recorded in the reporting. (4). Recording place is also not limited, but if possi- ble, please record somewhere quiet. (5). After every recording, please fill in a short survey about engagement (UWES-3).

Meanwhile, all participants were required to continuously participate in the video activity reporting trial for at least four weeks with no upper limit. After four weeks, they can quit the trial at any time.

3.3 Measures

In this trial, we used Utrecht Work Engagement Scale (UWES) [4] to measure participants' engagement [4]. Since it is a daily measurement, to not burden participants, we used the ultra-short version of UWES with only three questions (UWES-3) [28]. Followed by the idea of the student version of UWES (UWES-S) [15], we slightly rephrased the items and presented them as: (1). "When I'm doing my work as a student, I feel like I am bursting with energy (vigor)"; (2). "I am enthusiastic about my studies" (dedication); (3). "I am immersed in my studies (absorption)". Each item required a rate on a scale from 0 to 6, and the average of three ratings was calculated to present one's engagement level.

3.4 Results

Finally, we collected 232 video reports as well as the engagement surveys results in total. In detail, as shown in Table.1, the maximum number of video reports from participants is 56 while the minimum is 16. On the other hand, their actual trial days (except weekends and national holidays in Japan) distribute from 16 to 63.

4 Interview

After the trial, we conducted semi-structured interviews to in-depth investigate the feasibility of video activity reporting. Based on the interview with six trial participants, we explore the benefits and limitations of video activity reporting for students. Furthermore, we investigate how their characteristic behaviors in video activity reporting are influenced by engagement.

4.1 Participant

The participants for semi-structured interviews are from video activity reporting trial. We sent messages via SNS to all eight participants at a late date. As a result, six of them participated in this interview (except P5 and P7).

4.2 Procedure

The whole procedure took around an hour, including review and interview two sections. Since it is a post-day interview, participants were first required to look back on all their video reports in the review section, which can help them recall their memory during the trial period. The review session took 30-minutes around and was held by participants themselves. Afterward, a 30-minutes around interview session will conduct individually.

Due to the Covid-19, we conducted the interview session online via Zoom and recorded the whole process. The interview was conducted in English by two interviewers. Before the interview, the interviewers will check participants' video reports. During the process, interviewers will select several video reports with characteristic behaviors (e.g., missing recordings, engagement fluctuations, unusual actions, and frequently filled pauses/silent pauses).

The semi-structured interview focuses on two aspects: the evaluation of video activity reporting, and the relationship between participants' behaviors and engagement. After first asking participants' thoughts about reviewing their previous video reports, interviewers will in-depth explore the benefits and limitations of video activity reporting. Next, interviewers will share the selected video reports with participants through the screen share function in Zoom, and review them again with participants together. Combined with the reporting content, interviewers will dig out the reason for participants' behaviors and the connection with their engagement.

4.3 Findings

Afterward, all video-recorded interviews were transcribed. Based on the overall interviews, we elaborated on six frequently appeared themes: benefits and limitations in video activity reporting evaluation, as well as four kinds of characteristic behaviors, including missing recordings, engagement fluctuations, filled pauses and silent pauses, and unusual actions. Then we categorized participants' feedback to themes manually, and presented the summarized findings in this section.

Benefits of Video Activity Reporting. All six participants mentioned record such video activity reporting was helpful to them. Their commonly thought was video activity reporting provided them an opportunity to record daily progress and review previous progress. P1 stated, "I think recording such video report is good for me, which can let me record my daily progress. Especially when I was busy, it was hard for me to remember what I did last month. But the video report can be helpful".

Furthermore, three participants suggested keeping video activity reporting improved their summarization skills. They found their video reports become more concreted and structured in the late of the trial, which helps them better recognize their daily progress and improve self-efficiency.

On the other hand, two participants indicated that review video activity reporting could influence their engagement. P2 stated, "Before I start one day's recording, I always first review my yesterday's video and compare to confirm if there any progress I have made today. If there is, it makes me feel good, and I will be more engaged with the next work". It suggested video activity reporting helps participants recognize their achievements, which positively affects their engagement. On the contrary, review failures or bad progress can also bring negative effects. As P3 stated, "If things going well, it makes me feel good and more engaged, but when sometimes not, video reports let me recognize again about that, let me feel everything become worse".

Limitations of Video Activity Reporting. Students have difficulty in keeping the motivation of recording. Overall, four participants clearly stated video activity reporting burdens them and lets them feel bored. Regarding the reasons, frequent daily reporting in the trial might cause students' motivation to decrease. P8 stated, "I think it burdens me. Although a video report is only 30 seconds around, record every day made me feel boring and tried". Meanwhile, two participants mentioned that lack of feedback was another essential factor. Only reporting one side lets them feel recording is meaningless, resulting in difficulty maintaining motivation for the long term.

The differences between students and employees can make video activity reporting become harder for students. Compared with employees, students usually do not have the habit of reporting progress. It caused students to feel someone was forcing them to record and become less self-motived. P4 stated, "As I mentioned before, I think such video report is helpful to me. I think it's good. But if someone forces me to do it, I am not feeling good". Besides, academic activities can be harder to report than industry activities. As P1 stated, "It is hard for me to summarize my progress in daily reports.

Sometimes my research work cannot easily divide into daily tasks… And when I should record reports is another difficulty. Sometimes I work until midnight, which caused me cannot find a stable recording time". The research process is usually more complex than industry work, and academic work time is more unstable. These factors can also bring difficulties to students in daily reporting.

Reporting in video format is also one of the limitations. P4 stated, "I prefer the text report… When I write down something through text, it helps me to remember better. But if I record a video, it becomes more complicated because I need to use my computer or smartphone... Also, if I make any mistakes in the report, I must record it again. But if it is text, I can easily revise it". In practice, text activity reporting also has its convenience.

Missing Recordings. As shown in Table 1 above, five participants have experienced missing recordings during the trial. Regarding the reason, besides simply forgetting, four participants indicated they would intentionally didn't record when their progress was not ideal. P3 stated, "I didn't record that day just because my progress is not good". Furthermore, missing recordings can indicate one's lower engagement. P1 stated,"…that day I didn't forget it, but I just not engaged in my work, and caused I even don't want to record such a video report".

Engagement fluctuations. Participants' engagement scores sometimes suddenly changed, and we explored the reasons that caused such fluctuations. Five participants thought their research progress was the main factor. When they finish some tasks or get some achievements, their engagement usually becomes higher. P8 stated, "…I think because I finished many tasks that day, which makes me feel more engaged in my work". In contrast, when progress is not ideal, their engagement also significantly decreases. P6 stated, "Actually in that week the original plan is the task should finish in the week, so this day (Friday) I am focusing on it. But there are some problems, so I cannot finish on time. It means I need to work on weekends also to finish it, and that makes me feel less engaged."

In addition, three participants emphasized the role of the supervisor in their engagement changes. The comment and evaluations from their supervisor also significantly influence student's engagement. P1 stated, "When I submitted the work to my supervisor and got praised. It makes me feel more engaged in my work." While positive feedback improves one's engagement, negative feedback also has the opposite effects. P2 stated:" my supervisor asked me to redo one task, which means I have to work from the beginning. It makes me feel tired and causes me to give a low engagement score".

Filled Pauses and Silent Pauses. Filled pauses and silent pauses are two kinds of non-verbal cues that frequently occur in video activity reporting, and we explored the reason behind them. Three participants said filled pauses or silent pauses indicated they were considering. P4 stated, "Even I prepared the report content in my mind, I always forgot during the recording. Then I used filled pauses that trying to come up with the words". Also, P1 stated, "That day was hard to find what I can report during recording, so I frequently keep silent and thinking".

Unusual Actions. In video activity reporting, some of the participants' actions were observed as obviously different. These unusual actions sometimes can suggest considering and even indicate one's engagement. P8 frequently raised the head in one video report, which is unusual. Regarding the reason, P8 stated," I think I always raising my head is because I finished many tasks that day, and I wanted to recall them during the recording. Regarding the relationship to engagement, I frequently raised my head to remember because I finished many tasks that day. So finishing tasks makes me more engaged in my work".

5 Discussion

5.1 Qualitative Analysis

The qualitative analysis based on semi-structured interviews indicated that video activity reporting could bring benefits to students. It helps students better recognize their achievements and allows them to review previous progress. Since record daily highlights can bring significant positive outcomes to inner work life and worker wellbeing [11], we believe video activity reporting can also do that to students. Regularly record and review can improve students' awareness of progress recording and help them make plans, and long-term utilization additionally improves students' summarization skills. However, since students usually do not have the habit of activity recording, they might have difficulty adhering to long-term use. Apply systems that remind students to record and regular review as external supports can be solutions. Furthermore, to less experienced students, providing guidance on how to produce well-structured activity reporting can be helpful, maximizing the benefits to them.

The most significant barrier for students to video activity reporting is motivation. According to the interviews, the frequent daily recording makes participants feel bored and tired, resulting in low motivation. Meanwhile, complex academic research process increases the difficulty in daily reporting. Thus, rather than daily, less frequent activity reporting (e.g., weekly) might be more suitable for students. In addition, "reporting" usually refers to formal communication. It tends to make students feel that they are being forced and become less self-motivated. To students, informal journaling might be more acceptable than formal reporting.

Although we suggested video activity reporting to instead text one, video has its inconvenience in practice. Video requires specific devices and recording environments, while text doesn't have such limitations. In editing, video is also more complicated than text, which is hard to fix mistakes. Thus, combining video and text for activity reporting in practice is worth expecting to reduce the utilization burdens.

Our findings also demonstrated the potential for video activity reporting to provide engagement indicators. First, the recording itself can be an indicator of one's engagement. When one's progress is unsatisfied, and engagement decreased, the intention of recording activity reporting also declines. Therefore, deliberately miss recording can indicate declined engagement. Second, as most participants emphasized, their engagement is significantly influenced by progress. In video activity reporting, the unsatisfied progress can be indicated by word choice or specific cues according to individual habits. The indicators of unsatisfied progress can also suggest one's low engagement. Third, the feedback content from supervisors also has the potential to be an indicator of engagement. Positive and negative feedback content can influences one's engagement differently.

We also found changes in engagement can be reflected in the unusual actions in video activity reporting. However, such unusual actions are highly related to personal characteristics, so further analysis with more diverse data is needed to generalize the findings.

5.2 Limitations and Future Work

The participants for the video activity reporting trail are not diverse enough. Due to the limited participant size, how culture and gender influence the feasibility of video activity reporting remains unexplored.

On the other hand, the proposal of video activity reporting in [13] also suggested feedback from supervisors and managers. Many studies related to work engagement suggested feedback from managers can significantly influence one's engagement [32, 33, 34]. The feedback of video activity reporting can bring extra benefits of promoting remote workers' engagement. But in this study, the video activity reporting was only from students one-way, and what effects feedback can provide is still unexamined. Nevertheless, as participants mentioned in the interviews, without feedback significantly reduced their motivation. Due to the essential role of feedback in video activity reporting, we will explore its effects in our future works.

6 Conclusion

In this work, we investigated the feasibility of video activity reporting for university students. We first conducted a video activity reporting trial with eight university students. After four or more weeks of continuous daily recording of video reports, we collected 232 video reports and engagement survey answers. Afterward, we conducted semi-structured interviews on the video activity reporting with six participants. Our findings demonstrated the benefits and limitations of video activity reporting to students, while also explored the relation between their characteristic behaviors and engagement. Overall, video activity reporting can bring new opportunities for managing and measuring individual work engagement in remote work.

References

1. Raiborn, C., and Butler, J. B.: A new look at telecommuting and teleworking. Journal of Corporate Accounting & Finance 20(5), 31-39 (2009).
2. Sardeshmukh, S. R., Sharma, D., Golden, T. D.: Impact of telework on exhaustion and job engagement: A job demands and job resources model. New Technology, Work and Employment 27(3), 193-207 (2012).
3. Davis, R. and Cates, S.: The Dark Side of Working in a Virtual World: An Investigation of the Relationship between Workplace Isolation and Engagement among Teleworkers. Journal of Human Resource and Sustainability Studies 1, 9-13 (2013).
4. Schaufeli, W. B., et al.: The measurement of engagement and burnout: A two sample confirmatory factor analytic approach. Journal of Happiness studies 3(1), 71-92 (2002).
5. Harter, J. K., et al.: The relationship between engagement at work and organizational outcomes. Gallup Poll Consulting University Press, Washington (2013).
6. Jagannathan, A.: Determinants of employee engagement and their impact on employee performance. International Journal of Productivity and Performance Management 63(3), 308-323 (2014).
7. Jaupi, F., and Llaci, S.: Employee engagement and its relation with key economic indicators. Journal of Information Technology and Economic Development 5(2), 112-122 (2014).
8. Pogorilich, D. A.: The daily report as a job management tool. Cost Engineering 34(2), 23-25 (1992).
9. Etkina, E., and Harper, K. A.: Weekly reports: Student reflections on learning. Journal of College Science Teaching 31(7), 476-480 (2002).
10. Ito, K., Kizuka, A., and Oba, M. A.: Trial Utilization of Weekly Reports to Evaluate Learning for System Development PBLs. 5th IIAI International Congress on Advanced Applied Informatics (IIAI-AAI), pp. 1064-1067. IEEE, New York (2016).
11. Avrahami, D., et al.: Celebrating Everyday Success: Improving Engagement and Motivation using a System for Recording Daily Highlights. In: Proceedings of the 2020 CHI Conference on Human Factors in Computing Systems, pp. 1-13. ACM, New York (2020).
12. Lu, D., et al.: Challenges and Opportunities for Technology-Supported Activity Reporting in the Workplace. In: Proceedings of the 2018 CHI Conference on Human Factors in Computing Systems (2018).
13. He, Z., et al.: Preliminary Utility Study of a Short Video as a Daily Report in Teleworking. In: Proceedings 26th International Conference on Collaboration Technologies and Social Computing, pp. 35-49. Springer, Cham (2020).
14. He, Z., et al.: A Study of Speech Content in Daily video report for work engagement in teleworking (in Japanese). IPSJ Technical Report (GN) 2021.17, pp. 1-7. IPSJ, Tokyo (2021).
15. Schaufeli, W. B., et al.: Burnout and engagement in university students: A cross-national study. Journal of Cross-Cultural Psychology 33(5), 464-481 (2002).
16. Carmona-Halty, M. A., Schaufeli, W. B., and Salanova, M.: The Utrecht work engagement scale for students (UWES–9S): factorial validity, reliability, and measurement invariance in a Chilean sample of undergraduate university students. Frontiers in Psychology 10:1017, 5p. (2019).
17. Loscalzo, Y., and Giannini, M.: Study engagement in Italian university students: a confirmatory factor analysis of the Utrecht Work Engagement Scale—Student version. Social Indicators Research 142(2), 845-854 (2019).

18. Salanova, M., et al.: How obstacles and facilitators predict academic performance: The mediating role of study burnout and engagement. Anxiety, Stress, & Coping 23(1), 53-70 (2010).
19. Mitra, T., et al.: Spread of employee engagement in a large organizational network: A longitudinal analysis. In: Proceedings of the ACM on Human-Computer Interaction 1.CSCW pp. 1-20. ACM, New York (2017).
20. Soane, E., et al.: Development and application of a new measure of employee engagement: the ISA Engagement Scale. Human Resource Development International 15(5), 529-547 (2012).
21. Harter, J. K., Schmidt, F. L., and Hayes, T. L.: Business-unit-level relationship between employee satisfaction, employee engagement, and business outcomes: a meta-analysis. Journal of Applied Psychology 87(2), 268-279 (2002).
22. Rich, B. L., Lepine, J. A., and Crawford, E. R.: Job engagement: Antecedents and effects on job performance. Academy of Management Journal 53(3), 617-635 (2010).
23. Shami, N. Sadat, et al.: Inferring employee engagement from social media. In: Proceedings of the 33rd Annual ACM Conference on Human Factors in Computing Systems. ACM, NewYork (2015).
24. Kajiwara, Y., Shimauchi, T., and Kimura, H.: Predicting emotion and engagement of workers in order picking based on behavior and pulse waves acquired by wearable devices. Sensors 19(1), 165-186 (2019).
25. Kocielnik, R., et al.: Designing for workplace reflection: a chat and voice-based conversational agent. In: Proceedings of the 2018 Designing Interactive Systems Conference, pp. 881-894. ACM, New York (2018).
26. Bao, Lingfeng, et al.: Who will leave the company?: a large-scale industry study of developer turnover by mining monthly work report. 2017 IEEE/ACM 14th International Conference on Mining Software Repositories (MSR'17), pp. 170-181. IEEE, New York (2017).
27. Japanese-Language Proficiency Test, https://www.jlpt.jp/e/, last accessed 2021/09/10.
28. Schaufeli, W. B., et al.: An ultra-short measure for work engagement. European Journal of Psychological Assessment 35(4), 1-15 (2017).
29. Goto, M., Itou, K., and Hayamizu, S.: A real-time filled pause detection system for spontaneous speech recognition. Sixth European Conference on Speech Communication and Technology, pp. 227-230. ESCA, Budapest (1999).
30. Lee, M., et al.: Exploring moral conflicts in speech: multidisciplinary analysis of affect and stress. 2017 Seventh International Conference on Affective Computing and Intelligent Interaction (ACII), 8p. IEEE, New York (2017).
31. Delaborde, A., and Devillers, L.: Use of nonverbal speech cues in social interaction between human and robot: emotional and interactional markers. Proceedings of the 3rd International Workshop on Affective Interaction in Natural Environments, pp. 75–80. ACM, New York (2010).
32. Luthans, F., and Peterson, S. J.: Employee engagement and manager self-efficacy. Journal of management development 16(1), 57-72 (2002).
33. Donaldson-Feilder, E., and Lewis, R.: Positive manager behaviour for engagement and wellbeing. Flourishing in Life, Work and Careers. Edward Elgar Publishing, Cheltenham (2015).
34. Karanges, E, et al.: The influence of internal communication on employee engagement: A pilot study. Public Relations Review 41(1), 129-131 (2015).

An Experiment of Crowdsourced Online Collaborative Question Generation and Improvement for Video Learning materials in Higher Education

Ari Nugraha[1[0000-0002-5793-3157]] and Tomoo Inoue[2[0000-0003-3600-214X]]

[1] Graduate School of Library, Information and Media Studies, University of Tsukuba, Tsukuba, Japan
ari.nugraha@slis.tsukuba.ac.jp
[2] Faculty of Library, Information and Media Science, University of Tsukuba, Tsukuba, Japan
inoue@slis.tsukuba.ac.jp

Abstract. Learning by generating questions has proven to be an effective learning strategy. However, most of the studies on this topic involved students from a particular learning institution. In this study, we investigate how the learning activity of question generation and improvement can be implemented using a crowdsourcing platform. We use Amazon Mechanical Turk (Mturk) for crowd workers to create questions based on a video lecture and then improve those questions. Our main objective of this study is to study how question generation learning activity could improve learning gain and how question improvement by crowd workers could produce high-quality questions.

Keywords: Question Generation, Question Improvement, Video Learning Material, Crowdsourcing.

1 Introduction

Generating questions is a well-known learning strategy that facilitates students' cognitive and affective development. One of the main benefits offered by the question generation activity is that it helps students to identify the essential aspects of learning content and focus on its essential elements [1]. While most of the previous studies in this realm involved students from particular learning institution, in this study we want to explore how this learning activity is performed by crowd workers, particularly in a crowdsourcing platform. Crowdsourcing has been used in many domains, including in the domain of education. Educators and researchers around the world have been trying to use "the wisdom of crowd" or crowdsourcing to create educational content, practical experience, exchanging complementary knowledge, and augmenting feedbacks [2]. Not only as a learning strategy but in terms of creating educational content, one of the contents that can be produced with crowdsourcing is test question items. As a continuation of our previous study on online question improvement with undergraduate students [3], we will investigate how our design of question generation and improvement activity could improve the learning gain, and at the same time, we want to investigate how the iterative improvement process by the crowd workers could produce high-quality questions.

2 Literature Review

2.1 Question Generation

A recent study on question generation as a learning activity revealed that generating questions is a powerful learning strategy on par with testing on long-term retention and better than restudying [4]. The authors also claimed that the students who generated questions and answers performed similarly well as students who answered the authors' generated questions. The claim may imply that a student could create questions with a similar quality to a more experienced person such as a teacher. This also means that the questions generation not only can become a part of a learning activity for a student but at the same time can become a way for an educator to gather questions to be used for testing or learning assessment purposes. One example of this is that educators can utilize crowdsourcing platforms to generate question items for large-scale exams [5]. Efforts to automatically generate questions (automatic question generation/AQG) to overcome the challenge of a continuous supply of questions item have also been studied and reviewed where it has the potential to enable educators to spend more time on other important instructional activities [6].

In regards to the benefits of questions generation activity, it helps students to identify important parts of learning content in various forms, such as from a textbook [1] or a video [7]. In line with the study mentioned earlier, a study with underprepared college student shows that students who were trained to generate questions performed better compared to the students who wrote summarization and notetaking-reviewer in retention test of lecture content a week later [7]. Another benefit for students is that it promotes a higher level of thinking as the student is forced to seek the answer to their questions in the learning material [8]. However, it is also known that students have difficulties creating high-quality questions [9]. Students also felt that writing complex questions was burdensome and not trusting their peer-generated questions [10]. The low-quality questions produced by students or non-educator are understandable because producing high-quality question items require careful thinking about course learning outcomes and exploration of their own interpretations of particular concepts [11].

To improve the quality of a question, a question improvement activity is needed. An example of a strategy to improve the question is by providing scaffolding in the form of peer feedback[12]. Another collaborative way is by allowing another student to improve or change the question and then claim ownership of the question[13]. Our previous study with students in a synchronous online classroom shows that students can collaboratively improve their peer questions through an iterative improvement process[3, 14]. However, our studies also revealed that there are some issues with the improvement process such as students sometimes could not distinguish a low-level and a high-level question well which in turn made them difficult to improve their peers' questions.

In line with the previously mentioned studies [9–12], constant monitoring, and scaffolding through the improvement process from a more experienced person such as a teacher might be still needed to make sure students could pose a high-quality question.

2.2 Crowdsourcing as a Way to Generate Question

According to the Merriam-Webster dictionary, crowdsourcing is defined as the practice of obtaining needed services, ideas, or content by soliciting contributions from a large group of people and especially from the online community rather than from traditional employees or suppliers [15]. The anonymous people who work in crowdsourcing is usually called crowd worker. One of the successful examples from the Web 2.0 era of how voluntary online communities could create massive content is Wikipedia that at the time of this writing has 6,503,533 articles in English [16].

A common way to request a service from crowd workers is through a crowdsourcing platform. Several crowdsourcing platforms exist, such as Microworkers, Upwork, Innocentive, 99Design, Cad Crowd, CrowdSpring, and the most popular one, Amazon Mechanical Turk (Mturk). Some of these platforms are offering crowd workers for a specific type of job, while others are for more general crowd workers to do a simple microtask, such as filling out a survey, transcription of audio, or image classification. For research purposes, the Mturk platform has been widely known by researchers for data collection and study shows that workers in Mturk exhibit the classic heuristics and biases and pay attention to directions at least as much as subjects from traditional sources[17]. In the Mturk term, a task that needs to be done by a worker is called Human Intelligence Task (HIT).

A traditional way in the education field is that questions for student assessment are only generated by more experienced educators such as a teacher or faculty members. The Internet and Web have made online crowdsourcing a new way of obtaining services to create or generate ideas including educational content from many anonymous people on the Internet. A recent review of papers related to crowdsourcing in education revealed that the crowdsourcing method has been utilized in various types of learning activities such as personalized learning, homework, exam assessment, and learning material generation [18].

One of the early studies in an online crowdsourced question generation is PeerWise [19]. PeerWise is an online crowdsourcing tool that involves students in the process of creating, sharing, answering, and discussing multiple-choice questions (MCQ). It has been used by several studies in various fields of education such as computer programming [20], physics [21], medical [10, 22], and biology [23]. Evaluation of students' generated question quality in PeerWise shows that the students are capable of writing questions that faculty judge to be of high quality [20]. Crowdsourcing has also been used to create distractors for MCQ assessments and the result suggests that it was faster, easier, and cheaper than the traditional method [24].

3 Method

In this study, we use the Mturk platform to generate and improve questions according to the video lecture as learning material. The reason we used Mturk for this study is that it offers several features that are suitable for our studies such as accepting or rejecting task results from workers if it is not conformed with our instruction and the qualification feature that could be used to prescreen workers. As illustrated in Fig. 1, our crowdsourced activity comprises two phases, the first phase is question generation where we ask crowd workers to generate questions. The second phase is question improvement where we ask a series of tasks for different crowd workers to improve questions generated from the first phase.

3.1 Study Design

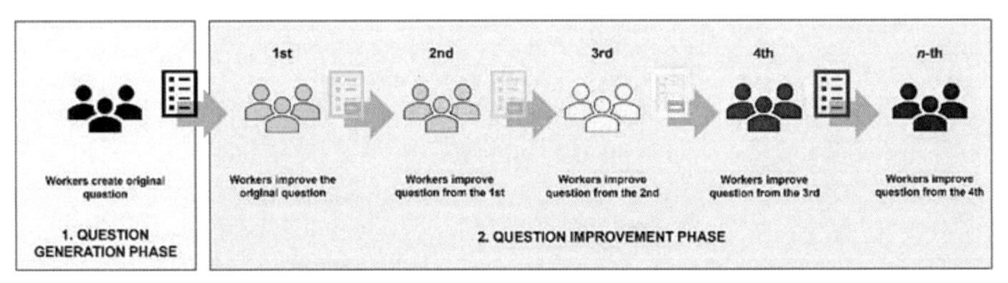

Fig. 1. Crowdsourced online collaborative question generation and improvement activity design in this study.

For this study, we use a between-subject design where we will compare the questions generated from the question generation phase and their subsequent (from the 1st until the nth iteration) improvements in the question improvement phase. Each iteration will be done by a different set of workers. Each worker in each iteration will have to generate five questions that will be improved by a different worker in the next iteration.

3.2 Learning Materials

Fig. 2. The candidate videos for our study are from the MIT OpenCourseWare channel on YouTube. The left side is a lecture video about Blockchain by Prof. Gary Gensler, and the right side is about Biology presented by Prof. Imperiali.

For learning materials as a source of question generation and improvement, we use lecture videos from video sharing site YouTube. The online lecture video that we used for this experiment is a short introductory video. The lecture video's duration should be around 15 minutes to allow participants to generate enough numbers of questions related to the learning content in the video. The reason we choose 15 minutes video is that workers will have many cognitive tasks such as answering pre-test, reading instructions, creating questions, and also re-watching the video while creating questions, which in total the expected duration will be 1 hour. Longer duration video will make the whole duration longer and might decrease worker motivation and time investment to do the task [25]. Although we used videos from YouTube, we defined several criteria for the lecture videos; we only used video lectures that have reusable licenses such as Creative Commons, presented by an authoritative or respected instructor/professor, and the instructor should appear in the video (example in Fig. 2) itself as it is considered to be more engaging [26].

Based on the criteria, one of our candidate videos is titled: *Introduction, Course Organization of MIT 7.016 Introductory Biology, Fall 2018* presented by Professors Imperiali of MIT (https://www.youtube.com/watch?v=KlVHqq38KJU). The video is taken from the MIT Open Courseware channel and published under the Creative Commons BY-NC-SA license. The original video duration is 38:45 minutes, and for this experiment, we take the 15 minutes general introduction content part of the video from the minutes 10:00 to the minutes 25:05 (the first 10 minutes of the original video only contains a class introduction to the professors and the general overview of the course itself). Other video candidates which have the same presentation style from MIT Open Courseware for this experiment are:

1. Introduction for Blockchain and Money (Introduction for 15.S12 Blockchain and Money, Fall 2018). Source: https://www.youtube.com/watch?v=EH6vE97qIP4
2. Cryptocurrency Engineering and Design: Signatures, Hashing, Hash Chains, e-cash, and Motivation (MIT MAS.S62 Cryptocurrency Engineering and Design, Spring 2018). Source: https://www.youtube.com/watch?v=IJquEYhiq_U

3.3 Pre-Test and Post-Test

To measure the learning gain of workers from question generation and improvement activity we have a pre-test and post-test section that participants need to answer. The questions for both tests will be the same and are based on the content of the video. The type of question that will be used for this experiment is a free recall question.

3.4 Worker Qualification

For this study, we target college/university student workers from 18 years old. Workers should also possess different educational backgrounds from the video content. We will recruit 10 workers for the question generation task and another 50 workers will be divided into five groups (10 workers/group) where each group will be assigned to each iteration in the question improvement phase. To make sure different unique workers participated in each iteration, we block the workers from the previous iteration(s) based on their Worker IDs. In this study, we only use the Worker IDs to block workers and not to identify them.

In our effort to make sure that only the workers who fall under our qualification can do the task, we created a custom qualification page that the worker needs to complete before they can work on our task. We also use the qualification page to inform consent about the experiment. We used the *Baseline Rewards* strategy for worker compensation [27]. Since our task requires a worker to do several sub-tasks in 60 minutes, the worker will be compensated $5.99, as the average hourly compensation for a task with certain qualifications is between $4.89 and $7.80 [28]. The worker will be paid if only they finish all the tasks and create the questions according to the instruction on the task page.

3.5 Task Design

Mturk provides two ways to build a HIT/task page, the first is using the Requester HIT editor and the second is programmatically using their SDK provided in several popular programming languages. For this study, we are using the Requester HIT editor to create our task page and then combined programmatically to create workers' qualifications using Python. By using only using the Requester HIT editor to build our task page, we simplify the data collection from workers as we do not need to build a separate external page. From the worker's side, they do not have to open an external site to complete the task.

Our task contains several sub-tasks where it is translated into several sections on the task page. Some of the sections will not open until workers submit or confirm that they have finished with that section. As pictured in Fig. 3, the pre-test section is visible the first time the worker enters our task page, but then it will be closed when they have submitted their pre-test answer. The pre-test submission will also trigger the page to open the "Watch a video lecture" section to allow the worker to watch the video. Our reason for the task page design is that we want the workers to focus on their sub-task before continuing to work on another sub-task.

For the improvement phase, the task design is similar but with a couple of notable differences as can be seen in Fig. 4, that it contains a new section called "Guidance to improve a question" which contains the information about the different levels of question and examples to improve the question. The level of questions is adapted from Guthrie's questioning rubric [29]. Another difference is that the question improvement task's page will contain other workers' questions from the previous iteration.

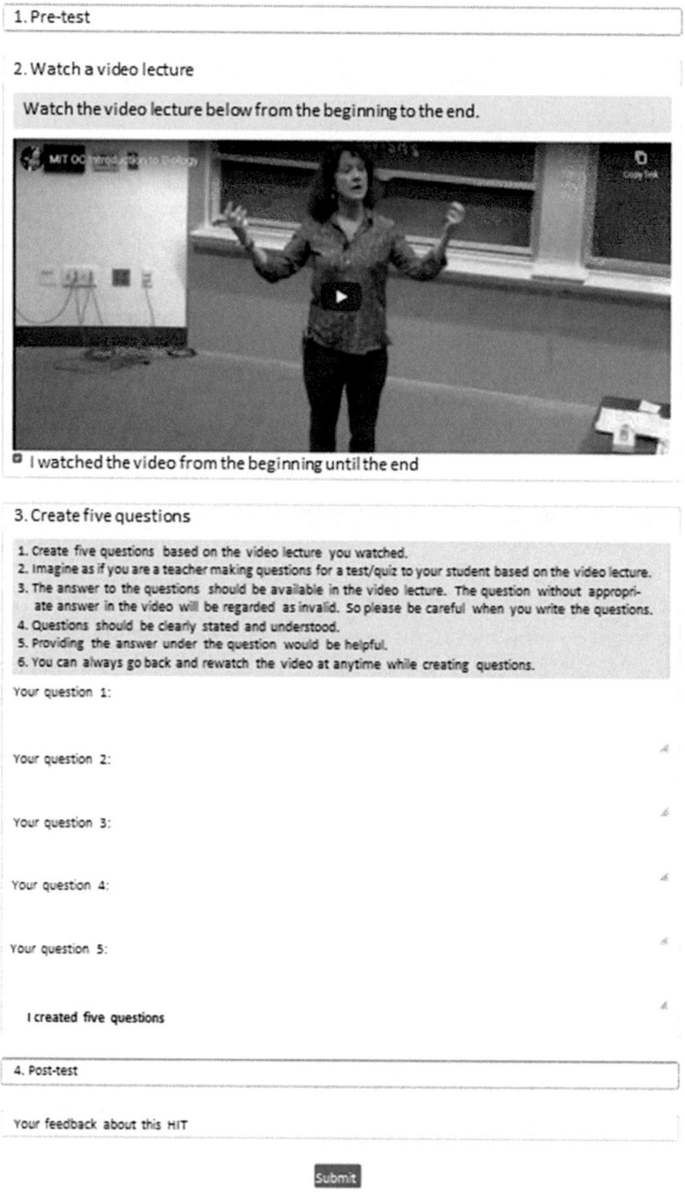

1. Pre-test

2. Watch a video lecture

Watch the video lecture below from the beginning to the end.

I watched the video from the beginning until the end

3. Create five questions

1. Create five questions based on the video lecture you watched.
2. Imagine as if you are a teacher making questions for a test/quiz to your student based on the video lecture.
3. The answer to the questions should be available in the video lecture. The question without appropriate answer in the video will be regarded as invalid. So please be careful when you write the questions.
4. Questions should be clearly stated and understood.
5. Providing the answer under the question would be helpful.
6. You can always go back and rewatch the video at anytime while creating questions.

Your question 1:

Your question 2:

Your question 3:

Your question 4:

Your question 5:

I created five questions

4. Post-test

Your feedback about this HIT

Submit

Fig. 3. HIT page design for the question generation phase contains a total of five sections: 1) Pre-test section, 2) Video (Watch a video lecture) section, 3) Question generation (Create five questions) section, 4) Post-test section, and 5) Feedback section. The lecture video is embedded into the task page to make it easier for a worker in reviewing the video while generating questions.

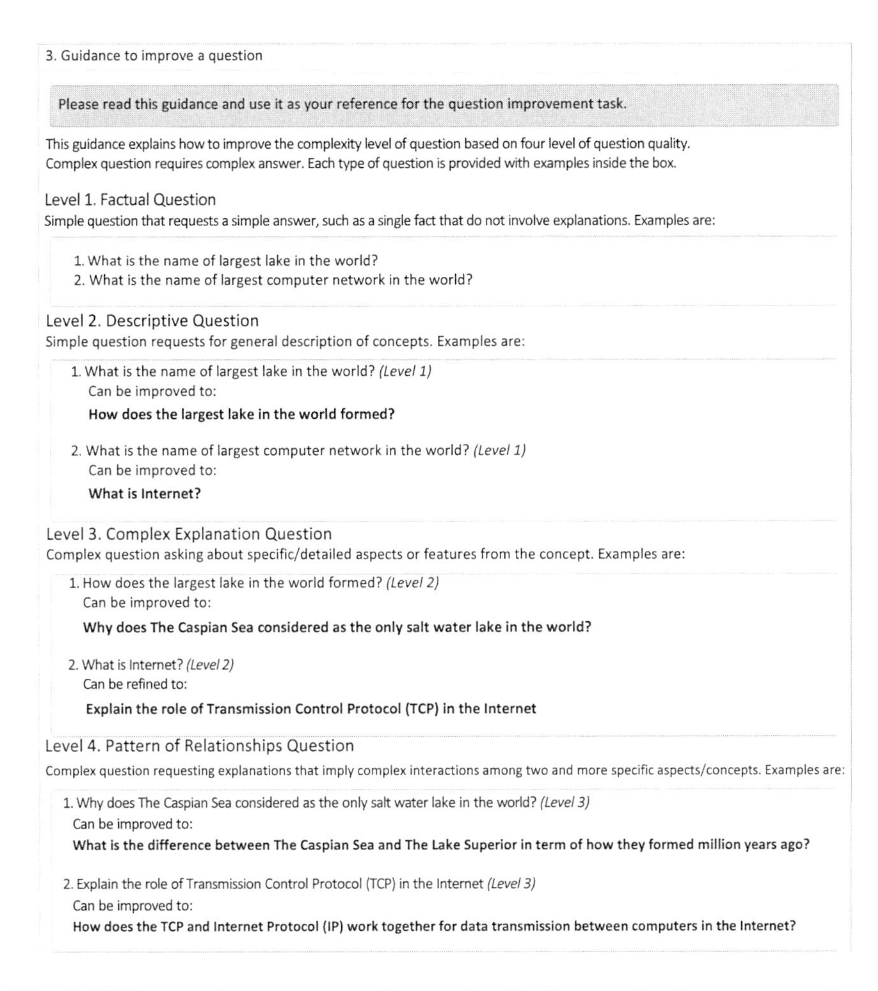
3. Guidance to improve a question

Please read this guidance and use it as your reference for the question improvement task.

This guidance explains how to improve the complexity level of question based on four level of question quality.
Complex question requires complex answer. Each type of question is provided with examples inside the box.

Level 1. Factual Question
Simple question that requests a simple answer, such as a single fact that do not involve explanations. Examples are:

1. What is the name of largest lake in the world?
2. What is the name of largest computer network in the world?

Level 2. Descriptive Question
Simple question requests for general description of concepts. Examples are:

1. What is the name of largest lake in the world? *(Level 1)*
   Can be improved to:
   How does the largest lake in the world formed?
2. What is the name of largest computer network in the world? *(Level 1)*
   Can be improved to:
   What is Internet?

Level 3. Complex Explanation Question
Complex question asking about specific/detailed aspects or features from the concept. Examples are:

1. How does the largest lake in the world formed? *(Level 2)*
   Can be improved to:
   Why does The Caspian Sea considered as the only salt water lake in the world?
2. What is Internet? *(Level 2)*
   Can be refined to:
   Explain the role of Transmission Control Protocol (TCP) in the Internet

Level 4. Pattern of Relationships Question
Complex question requesting explanations that imply complex interactions among two and more specific aspects/concepts. Examples are:

1. Why does The Caspian Sea considered as the only salt water lake in the world? *(Level 3)*
   Can be improved to:
   What is the difference between The Caspian Sea and The Lake Superior in term of how they formed million years ago?
2. Explain the role of Transmission Control Protocol (TCP) in the Internet *(Level 3)*
   Can be improved to:
   How does the TCP and Internet Protocol (IP) work together for data transmission between computers in the Internet?

Fig. 4. Guidance to improve a question section for the question improvement phase

3.6 Worker Task Workflow

In this section, we describe the workflow that the worker needs to do to complete our task.

1. After we publish a batch of task to Mturk, a worker can see our published task page at https://worker.mturk.com/, and if they are interested in the task, they need to take a qualification test first which also contains the consent form.
2. After completing the qualification and agreeing with the consent form, the worker can then press the “Accept & Work” button to start the task.
3. Upon entering the task page, the worker needs to work on a short pre-test by answering five questions provided. The worker then needs to click on the "Save answer and watch the video" button if they are finished with the pre-test. Once

the worker clicks the button, the pre-test section block will be closed, and the worker could not write the answer anymore.

4. The video section block will be opened to allow the worker to watch the video lecture. Once they are finished with the video, they must tick on the "I watched the video from the beginning until the end" checkbox to confirm that they have watched the whole video.
5. After the worker finishes watching the video, they need to create questions related to the learning content in the video inside the provided input boxes in creating questions section. To make sure and remind the worker about the question generation tasks, the worker then must tick the "I created five questions" checkbox.
6. Answer the post-test questions.
7. Writes feedback about the whole experience with the task.
8. Submits the task by pressing the Submit button at the bottom of the page.

4 Data Analysis

Based on the result that we will gather from the crowd workers, two main types of data that we will use for analysis. The first is the questions data generated from workers and the second one is the workers' answers for pre-test and post-test. Our main analysis for the questions data will be into the quality of questions, which we will rate based on the levels from Guthrie's questioning rubric [29]. We will also look into how the content of the questions evolved through iterations of improvement. The second analysis is from the scoring of pre-test and post-test answers to measure how much learning was gained by the question generation and improvement activity.

5 Expected Result

From this study we expected to have high-quality questions generated and improved by the crowd workers and that the questions will have an incremental increase in their transformation across iterations in term of quality and content.

References

1. Rosenshine, B., Meister, C., Chapman, S.: Teaching Students to Generate Questions: A Review of the Intervention Studies. Review of Educational Research. 66, 181–221 (1996). https://doi.org/10.3102/00346543066002181
2. Jiang, Y., Schlagwein, D., Benatallah, B.: A Review on Crowdsourcing for Education: State of the Art of Literature and Practice. In: Proceedings of The 22nd Pacific Asia Conference on Information Systems (PACIS 2018). p. 180 (2018)
3. Nugraha, A., Inoue, T.: Improving Students' Question Quality Through Online Iterative Refinement Activity. In: Bastiaens, T. (ed.) Proceedings of Innovate Learning Summit

2021. pp. 186–196. United States: Association for the Advancement of Computing in Education (AACE) (2021)
4. Ebersbach, M., Feierabend, M., Nazari, K.B.B.: Comparing the effects of generating questions, testing, and restudying on students' long-term recall in university learning. Appl Cognit Psychol. 34, 724–736 (2020). https://doi.org/10.1002/acp.3639
5. Alghamdi, E.A., Aljohani, N.R., Alsaleh, A.N., Bedewi, W., Basheri, M.: CrowdyQ: a virtual crowdsourcing platform for question items development in higher education. In: Proceedings of the 17th International Conference on Information Integration and Web-based Applications & Services. pp. 1–4. ACM, Brussels Belgium (2015)
6. Kurdi, G., Leo, J., Parsia, B., Sattler, U., Al-Emari, S.: A Systematic Review of Automatic Question Generation for Educational Purposes. Int J Artif Intell Educ. 30, 121–204 (2020). https://doi.org/10.1007/s40593-019-00186-y
7. King, A.: Comparison of Self-Questioning, Summarizing, and Notetaking-Review as Strategies for Learning From Lectures. American Educational Research Journal. 29, 303–323 (1992). https://doi.org/10.3102/00028312029002303
8. Papinczak, T., Peterson, R., Babri, A.S., Ward, K., Kippers, V., Wilkinson, D.: Using student-generated questions for student-centred assessment. Assessment & Evaluation in Higher Education. 37, 439–452 (2012). https://doi.org/10.1080/02602938.2010.538666
9. Logtenberg, A., van Boxtel, C., van Hout-Wolters, B.: Stimulating situational interest and student questioning through three types of historical introductory texts. Eur J Psychol Educ. 26, 179–198 (2011). https://doi.org/10.1007/s10212-010-0041-6
10. Grainger, R., Dai, W., Osborne, E., Kenwright, D.: Medical students create multiple-choice questions for learning in pathology education: a pilot study. BMC Med Educ. 18, 201 (2018). https://doi.org/10.1186/s12909-018-1312-1
11. Hancock, D., Hare, N., Denny, P., Denyer, G.: Improving large class performance and engagement through student-generated question banks. Biochemistry and Molecular Biology Education. 46, (2018). https://doi.org/10.1002/bmb.21119
12. Yu, F.-Y.: Multiple peer-assessment modes to augment online student question-generation processes. Computers & Education. 56, 484–494 (2011). https://doi.org/10.1016/j.compedu.2010.08.025
13. Yeckehzaare, I., Barghi, T., Resnick, P.: QMaps: Engaging Students in Voluntary Question Generation and Linking. In: Proceedings of the 2020 CHI Conference on Human Factors in Computing Systems. pp. 1–14. ACM, Honolulu HI USA (2020)
14. Nugraha, A., Wahono, I.A., Inoue, T.: An Initial Study of Collaborative Refinement of Student-Created Questions. In: Conference Companion Publication of the 2020 on Computer Supported Cooperative Work and Social Computing. pp. 359–363. Association for Computing Machinery, New York, NY, USA (2020)
15. Definition of CROWDSOURCING, https://www.merriam-webster.com/dictionary/crowdsourcing
16. Main Page, https://en.wikipedia.org/w/index.php?title=Main_Page&oldid=1085170884, (2022)
17. Paolacci, G.: Running experiments on Amazon Mechanical Turk. Judgment and Decision Making 5, 9 (2010)
18. Alenezi, H.S., Faisal, M.H.: Utilizing crowdsourcing and machine learning in education: Literature review. Educ Inf Technol. 25, 2971–2986 (2020).

https://doi.org/10.1007/s10639-020-10102-w

19. Denny, P., Luxton-Reilly, A., Hamer, J.: The PeerWise System of Student Contributed Assessment Questions. 78, 6 (2008)
20. Denny, P., Luxton-Reilly, A., Simon, B.: Quality of student contributed questions using PeerWise. In: Proceedings of the Eleventh Australasian Conference on Computing Education. pp. 55–63. Australian Computer Society, Inc. (2009)
21. Bates, S.P., Galloway, R.K., Homer, D., Riise, J.: Assessing the quality of a student-generated question repository. Phys. Rev. ST Phys. Educ. Res. 10, 020105 (2014). https://doi.org/10.1103/PhysRevSTPER.10.020105
22. Tackett, S., Raymond, M., Desai, R., Haist, S.A., Morales, A., Gaglani, S., Clyman, S.G.: Crowdsourcing for assessment items to support adaptive learning. Med Teach. 40(8), 838–841 (2018). https://doi.org/10.1080/0142159X.2018.1490704
23. McQueen, H.A., Shields, C., Finnegan, D.J., Higham, J., Simmen, M.W.: Peerwise provides significant academic benefits to biological science students across diverse learning tasks, but with minimal instructor intervention: Diverse Learning Benefits of PeerWise for Biology Students. Biochem Mol Biol Educ. 42(5), 371–381 (2014). https://doi.org/10.1002/bmb.20806
24. Scheponik, T., Golaszewski, E., Herman, G., Offenberger, S., Oliva, L., Peterson, P.A.H., Sherman, A.T.: Investigating Crowdsourcing to Generate Distractors for Multiple-Choice Assessments. In: Choo, K.-K.R., Morris, T.H., and Peterson, G.L. (eds.) National Cyber Summit (NCS) Research Track. pp. 185–201. Springer International Publishing, Cham (2020)
25. Yang, K., Qi, H.: The Nonlinear Impact of Task Rewards and Duration on Solvers' Participation Behavior: A Study on Online Crowdsourcing Platform. JTAER. 16, 709–726 (2021). https://doi.org/10.3390/jtaer16040041
26. Guo, P.J., Kim, J., Rubin, R.: How video production affects student engagement. Proceedings of the first ACM conference on Learning @ scale conference - L@S '14. 41–50 (2014). https://doi.org/10.1145/2556325.2566239
27. Chen, J.J., Menezes, N.J., Bradley, A.D.: Opportunities for Crowdsourcing Research on Amazon Mechanical Turk. 5
28. Hara, K., Adams, A., Milland, K., Savage, S., Callison-Burch, C., Bigham, J.P.: A Data-Driven Analysis of Workers' Earnings on Amazon Mechanical Turk. In: Proceedings of the 2018 CHI Conference on Human Factors in Computing Systems. pp. 1–14. ACM, Montreal QC Canada (2018)
29. Guthrie, J.T., Wigfield, A., Perencevich, K.C. eds: Motivating reading comprehension: concept-oriented reading instruction. L. Erlbaum Associates, Mahwah, N.J (2004)

Compact N-gram Language Models for Armenian

Davit Karamyan[1[0000−0002−6718−1593]], Ara Abovyan[2[0000−0002−1030−9437]], and Tigran Karamyan[3[0000−0002−0449−4920]]

[1] Russian-Armenian University, Yerevan, Armenia
[2] American University of Armenia, Yerevan, Armenia
[3] Yerevan State University, Yerevan, Armenia
davitkar98@gmail.com

Abstract. Applications such as speech recognition and machine translation use language models to select the most likely translation among many hypotheses. For on-device applications, inference time and model size are just as important as performance. In this work, we explored the fastest family of language models: the N-gram models for the Armenian language. In addition, we researched the impact of pruning and quantization methods on model size reduction. Finally, we used Bye Pair Encoding (BPE) to build a subword language model. As a result, we obtained a compact (100 MB) subword language model trained on massive Armenian corpora.

Keywords: Armenian l anguage, N-gram Language Model, Subword Language Model, Pruning, Quantization.

1 Introduction

Language modeling is a fundamental task of natural language processing (NLP). Models that assign probabilities to sequences of tokens are called language models or LMs. Here, tokens can be either words, characters, or subwords. The N-gram is the simplest model that assigns probabilities to sentences and sequences of tokens. Although the N-gram models are much simpler than modern neural language models based on recurrent neural network (RNN) [11, 19] and transformers [1, 4, 18], they are much faster than others since they perform constant-time lookups and scalar multiplications (instead of matrix multiplications in neural models). As always, trade-offs exist between time, space, and accuracy [2]. Hence, much recent work has been looking at building faster and smaller N-gram language models [5, 6, 17].

N-gram language models are widely utilized in spelling correction [10], speech recognition [7] and machine translation [21] systems. In such systems, for each utterance/sentence translation, the system generates several alternative token sequences and scores them using N-gram LM to peek the most likely translation sequence. In addition, LM rescoring can be combined with beam search algorithms [8].

Armenian language has a rich morphology: one word can have several tenses and surface forms. Moreover, in the Armenian language, one can form long words by stringing together word pieces. The inclusion of every form in the vocabulary will make it intractable. Subword dictionaries, in which words are divided into frequent parts, can help reduce vocabulary size. There have been many efforts in using word decomposition and subword LMs for dealing with out-of-vocabulary words in inflective languages such as Arabic [15], Finnish [23], Russian [16], and Turkish [24]. Review of literature revealed that there have been no publicly available LM resources for the Armenian language. This work is devoted to creating a compact and fast N-gram LM for the Armenian language.

In summary, we will give answers to the following practical questions: **Q1**. What order of N-grams is enough to build a good LM for the Armenian language? **Q2**. How much data is needed to build a model? **Q3**. How can pruning and quantization help reduce the size of the model? **Q4**. Can we build more compact models by using subwords?

In addition, we are going to release training codes and models.[4]

2 Background

Language Modeling (LM) is the task of predicting what token or word comes next. You might also think of an LM as a system that assigns probability to a piece of text. The probability of a sequence of n tokens $t_1^n := \{t_1, ..., t_n\}$ is denoted as $P(t_1^n)$. Using the chain rule of probability we can decompose this probability:

$$P(\{t_1, ..., t_n\}) = \prod_{k=1}^{n} P(t_k|t_1^{k-1})$$

Instead of computing the probability of a token given its entire history, it is usually conditioned on a window of N previous tokens. The assumption that the probability of a token depends only on the previous $N-1$ token is called a Markov assumption:

$$P(t_k|t_1^{k-1}) \approx P(t_k|t_{k-N+1}^{k-1})$$

We can estimate the probabilities of an N-gram model by getting counts from a corpus, and normalizing the counts so that they lie between 0 and 1. For example, to compute a particular N-gram probability of a token t_k given a previous tokens t_{k-N+1}^{k-1}, we'll compute the count of the N-gram t_{k-N+1}^{k} and normalize by the sum of all the N-grams that share the same prefix t_{k-N+1}^{k-1}:

$$P(t_k|t_{k-N+1}^{k-1}) = \frac{Count(t_{k-N+1}^{k})}{\sum_t Count(t_{k-N+1}^{k-1}, t)} = \frac{Count(t_{k-N+1}^{k})}{Count(t_{k-N+1}^{k-1})}$$

where $Count(x)$ is the number of times the token sequence x appears in the training corpus.

[4] https://github.com/naymaraq/arm-n-gram

There are two major problems with N-gram language models: storage and sparsity. To compute N-gram probabilities we need to store counts for all N-grams in the corpus. As N increases or the corpus size increases, the model size increases as well. Pruning and Quantization may provide a partial solution to reduce the model size. Any N-gram that appeared a sufficient number of times might have a reasonable estimate for its probability. But because any corpus is limited, some perfectly acceptable tokens may never appear in the corpus. As a result of it, for any training corpus, there will be a substantial number of cases of putative "zero probability N-grams". To keep an LM from assigning zero probability to these unseen events, we'll have to shave off a bit of probability mass from some more frequent events and give it to the events we've never seen. This is called smoothing. There are many ways to do smoothing: add-one(add-k) smoothing, backoff, and Kneser-Ney smoothing [12].

3 Experiments

3.1 Setup

We estimate N-gram probabilities on Armenian Wikipedia corpus[5] and CC-100 Web Crawl Data[6] [3]. To test the language models, we compute perplexity on two test datasets: Armenian Paraphrase Detection Corpus[7] (ARPA [14]) and Universal Dependencies treebank[8] (UD). The perplexity of a language model can be understood as a measure of uncertainty when predicting the next token.

All datasets are normalized by removing punctuation marks and non-Armenian symbols. Table 1 provides some statistics of the data after all normalization steps have been performed. Table 2 shows unique N-gram counts presented in training corpus.

We are going to measure perplexity of the corpus C that contains m sentences and N tokens. Let's the sentences $(s_1, s_2, ..., s_m)$ be part of C. Under assumption that those sentences are independent, the perplexity of the corpus is given by:

$$Perp(C) = \sqrt[N]{\frac{1}{p(s_1, s_2, ..., s_m)}} = \sqrt[N]{\frac{1}{\prod_{k=1}^{m} p(s_k)}}$$

We use KenLM [9] to train language models. KenLM implements two data structures: Probing and Trie, for efficient language model queries, reducing both time and memory costs. KenLM estimates language model parameters from text using modified Kneser-Ney smoothing.

[5] https://github.com/YerevaNN/word2vec-armenian-wiki
[6] https://data.statmt.org/cc-100/
[7] https://github.com/ivannikov-lab/arpa-paraphrase-corpus
[8] https://github.com/UniversalDependencies/UD_Armenian-ArmTDP

Table 1. Datasets statistics.

Dataset	Tokens (M)	Bytes	Split
CC-100	409	5.4Gb	train
Wiki	18.6	249Mb	train
ARPA	0.133	1.8Mb	test
UD	0.034	425Kb	test

Table 2. N-gram counts.

Order (N)	Count of unique N-grams
1	3648574
2	60190581
3	160796455
4	217396323
5	233510708

3.2 Q1. Order of Grams vs Perplexity

To determine what order of N-grams is sufficient to build a good LM for Armenian, we trained several LMs with different orders and calculated perplexity on the test datasets. Figure 1 shows the trend between perplexity and order of N-gram. It also shows how the size of the model changes as N increases.

From Figure 1 we can deduce that the effective orders are 5 and 6 grams. Although their sizes are quite large: 3.9GB and 5.5GB.

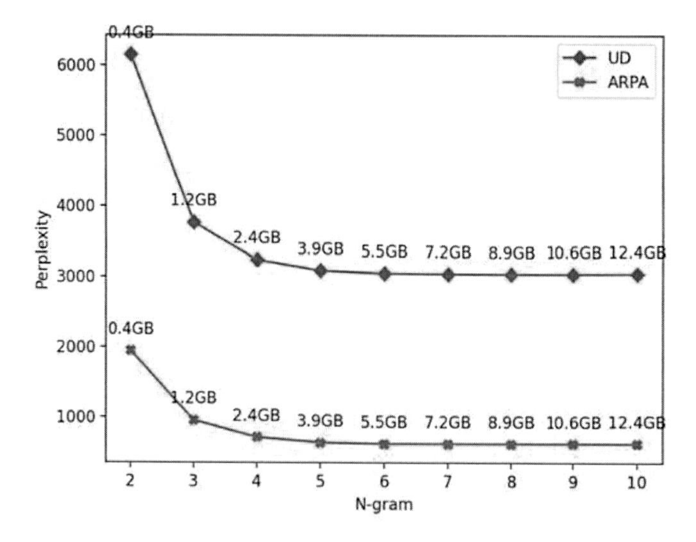

Fig. 1. N-gram order vs perplexity.

3.3 Q2. Training Corpus Size vs Perplexity

The next question we would like to ask is about corpus size. If the training corpus is small, we will end up with a very sparse model, and all perfectly acceptable Armenian tokens will be considered unknown. To find out how much

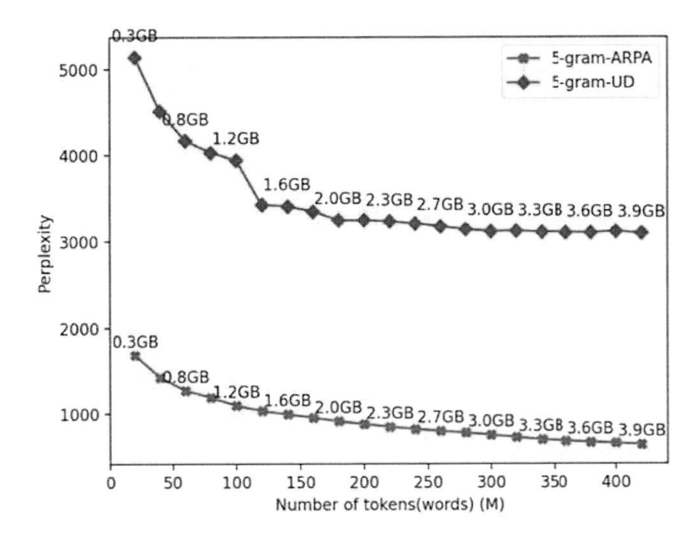

Fig. 2. Number of tokens in training corpus vs perplexity.

data is required, we shuffled and divided the entire training corpus into parts and trained a 5-gram LM for each part. Figure 2 shows the trend between perplexity and corpus size.

It can be seen that perplexity reaches saturation when the number of tokens exceeds 380M. Of course, there is always a trade-off between corpus size, perplexity and the model size: the larger the corpus size, the less perplexity and the larger the model.

3.4 Q3. Quantization and Pruning

On-device applications should be as compact as possible. So, the next question we would like to raise concerns the size of the model. Can we build a smaller LM without sacrificing performance?

To reduce the size of the model, we prune all n-grams that appear in the training corpus less than or equal to a given threshold. In addition, we use quantized probabilities by setting fewer bits. For this experiment, we trained a 5-gram LM.

The effect of pruning and quantization is provided in Table 3. Quantization can help reduce the size of a model by a couple of megabytes without perplexity degradation. In contrast, pruning drastically reduces the size of the model at the cost of worsening perplexity. For example, removing all n-grams less than or equal to 4 can reduce the size of the model by more than 12 times with a relative perplexity degradation of 36% for the UD dataset and 100% for the ARPA dataset.

Table 3. The effect of pruning and quantization on the trade-off between size and perplexity.

Pruning threshold	N-bits	Size	UD	ARPA
0	5	3.44Gb	3043.47	631.58
0	6	3.59Gb	3068.62	638.84
0	7	3.74Gb	3075.99	641.57
0	8	3.9Gb	3089.41	642.93
2	5	481.28Mb	3781.29	1131.82
2	6	501.76Mb	3768.36	1128.14
2	7	512.0Mb	3767.81	1125.54
2	8	532.48Mb	3764.69	1125.0
4	5	296.96Mb	4252.71	1344.56
4	6	307.2Mb	4219.03	1335.89
4	7	317.44Mb	4218.13	1332.73
4	8	317.44Mb	4217.73	1332.84
6	5	245.76Mb	4473.19	1486.95
6	6	245.76Mb	4432.03	1474.89
6	7	256.0Mb	4435.29	1471.75
6	8	256.0Mb	4431.23	1471.89
8	5	215.04Mb	4694.73	1588.09
8	6	225.28Mb	4655.29	1576.21
8	7	225.28Mb	4652.95	1571.46
8	8	225.28Mb	4652.89	1571.94

3.5 Q4. Subword Language Modeling

So far, we have considered text as a sequence of words separated by a space. Space tokenization is an example of word tokenization, which is defined as breaking sentences into words. Word tokenization method can lead to problems for massive text corpora and usually generates a very big vocabulary (e.g. our training corpus contains $3,648,574$ unique tokens, see Table 1). Instead of using word tokenization, we will use subword tokenization, which is based on the principle that frequently used words should not be split into smaller subwords, but rare words should be decomposed into meaningful subwords. There are several subword tokenization algorithms: Byte-Pair Encoding [22] , WordPiece [20], and SentencePiece [13]. Subword tokenization allows the model to have a reasonable vocabulary size. In addition, subword tokenization enables the model to process words it has never seen before, by decomposing them into known subwords. This is especially useful in agglutinative languages such as Armenian, where you can form long words by stringing together subwords.

We trained a BPE tokenizer with a vocabulary size of 128 using the SentencePiece package[9]. Next, we build several N-gram models on a tokenized corpus. Figure 3 shows the trend between perplexity and order of N-gram for subword model. It also shows how the size of the model changes as N increases.

[9] https://github.com/google/sentencepiece

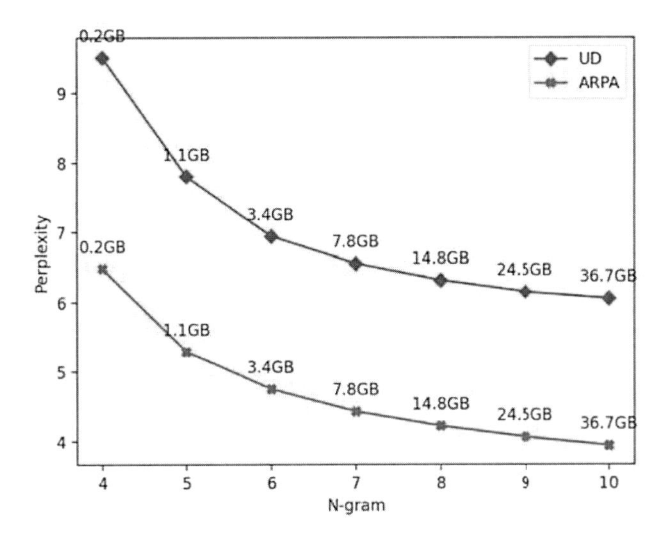

Fig. 3. N-gram order vs perplexity (subword).

Table 4. Pruning effect for the subword model with 10-gram.

Pruning	Size	UD	ARPA
0	36.66Gb	6.055	3.941
2	1.11Gb	6.199	4.19
4	634.88Mb	6.323	4.306
6	440.32Mb	6.373	4.381
8	348.16Mb	6.435	4.44
10	286.72Mb	6.53	4.491
16	184.32Mb	6.781	4.619
20	153.6Mb	6.892	4.69
24	122.88Mb	7.02	4.751
30	102.4Mb	7.146	4.837

First, in the Figure 3 the perplexity (0-10) is significantly lower than perplexity of the word-based tokenized model (0-7000, see Fig. 1). This is because we no longer have unknown tokens. In contrast to word-based models, subword models are much larger (e.g. 10-gram subword model is 3 times bigger).

Since the sequences no longer contain words, but contain subwords, in order to capture sufficient context, we need to consider higher order grams. From the figure 3 it can be seen that the higher the order, the larger the model (for example, a subword model with 10-gram has a size of 36.7 GB). To reduce the size of the model, we use pruning again. Table 4 provides information about the pruning effect for the subword model with 10-gram. It can be seen that we can reduce the model size by a factor of 368 from 36.7 GB to 102 MB with a relative perplexity degradation of 18% for the UD dataset and 23% for the ARPA dataset.

4 Conclusions

In this article, we have explored N-gram language models for the Armenian language. Our experiments have shown that for word-based language models, the effective orders are 5 and 6. In contrast, the effective order for subword language models can be higher than 10.

Also, we have explored the impact of pruning and quantization on the trade-off between model size and perplexity. Quantization can help reduce the size of the model without significantly degrading perplexity. Pruning, on the other hand, drastically reduces the size of the model at the expense of aggravating perplexity. For the subword language model, the perplexity degradation is much lower than for the word-based language model.

We have released compact N-gram language models built on very large corpora.

References

1. Brown, T.B., Mann, B., Ryder, N., Subbiah, M., Kaplan, J., Dhariwal, P., Neelakantan, A., Shyam, P., Sastry, G., Askell, A., et al.: Language models are few-shot learners. arXiv preprint arXiv:2005.14165 (2020)
2. Buck, C., Heafield, K., Van Ooyen, B.: N-gram counts and language models from the common crawl. In: LREC. vol. 2, p. 4. Citeseer (2014)
3. Conneau, A., Khandelwal, K., Goyal, N., Chaudhary, V., Wenzek, G., Guzmán, F., Grave, E., Ott, M., Zettlemoyer, L., Stoyanov, V.: Unsupervised cross-lingual representation learning at scale. arXiv preprint arXiv:1911.02116 (2019)
4. Devlin, J., Chang, M.W., Lee, K., Toutanova, K.: Bert: Pre-training of deep bidirectional transformers for language understanding. arXiv preprint arXiv:1810.04805 (2018)
5. Germann, U., Joanis, E., Larkin, S.: Tightly packed tries: How to fit large models into memory, and make them load fast, too. In: Proceedings of the Workshop on Software Engineering, Testing, and Quality Assurance for Natural Language Processing (SETQA-NLP 2009). pp. 31–39 (2009)
6. Guthrie, D., Hepple, M.: Storing the web in memory: Space efficient language models with constant time retrieval. In: Proceedings of the 2010 Conference on Empirical Methods in Natural Language Processing. pp. 262–272 (2010)
7. Hannun, A., Case, C., Casper, J., Catanzaro, B., Diamos, G., Elsen, E., Prenger, R., Satheesh, S., Sengupta, S., Coates, A., et al.: Deep speech: Scaling up end-to-end speech recognition. arXiv preprint arXiv:1412.5567 (2014)
8. Hannun, A.Y., Maas, A.L., Jurafsky, D., Ng, A.Y.: First-pass large vocabulary continuous speech recognition using bi-directional recurrent dnns. arXiv preprint arXiv:1408.2873 (2014)
9. Heafield, K.: Kenlm: Faster and smaller language model queries. In: Proceedings of the sixth workshop on statistical machine translation. pp. 187–197 (2011)

10. Hernandez, S.D., Calvo, H.: Conll 2014 shared task: Grammatical error correction with a syntactic n-gram language model from a big corpora. In: Proceedings of the Eighteenth Conference on Computational Natural Language Learning: Shared Task. pp. 53–59 (2014)
11. Hochreiter, S., Schmidhuber, J.: Long short-term memory. Neural computation **9**(8), 1735–1780 (1997)
12. Jurafsky, D.: Speech & language processing. Pearson Education India (2000)
13. Kudo, T., Richardson, J.: Sentencepiece: A simple and language independent subword tokenizer and detokenizer for neural text processing. arXiv preprint arXiv:1808.06226 (2018)
14. Malajyan, A., Avetisyan, K., Ghukasyan, T.: Arpa: Armenian paraphrase detection corpus and models. In: 2020 Ivannikov Memorial Workshop (IVMEM). pp. 35–39. IEEE (2020)
15. Mousa, A.E.D., Kuo, H.K.J., Mangu, L., Soltau, H.: Morpheme-based feature-rich language models using deep neural networks for lvcsr of egyptian arabic. In: 2013 IEEE International Conference on Acoustics, Speech and Signal Processing. pp. 8435–8439. IEEE (2013)
16. Oparin, I.: Language models for automatic speech recognition of inflectional languages. Dizertační práce, University of West Bohemia (2008)
17. Pauls, A., Klein, D.: Faster and smaller n-gram language models. In: Proceedings of the 49th annual meeting of the Association for Computational Linguistics: Human Language Technologies. pp. 258–267 (2011)
18. Raffel, C., Shazeer, N., Roberts, A., Lee, K., Narang, S., Matena, M., Zhou, Y., Li, W., Liu, P.J.: Exploring the limits of transfer learning with a unified text-to-text transformer. arXiv preprint arXiv:1910.10683 (2019)
19. Sarzynska-Wawer, J., Wawer, A., Pawlak, A., Szymanowska, J., Stefaniak, I., Jarkiewicz, M., Okruszek, L.: Detecting formal thought disorder by deep contextualized word representations. Psychiatry Research **304**, 114135 (2021)
20. Schuster, M., Nakajima, K.: Japanese and korean voice search. In: 2012 IEEE International Conference on Acoustics, Speech and Signal Processing (ICASSP). pp. 5149–5152. IEEE (2012)
21. Schwenk, H., Déchelotte, D., Gauvain, J.L.: Continuous space language models for statistical machine translation. In: Proceedings of the COLING/ACL 2006 Main Conference Poster Sessions. pp. 723–730 (2006)
22. Sennrich, R., Haddow, B., Birch, A.: Neural machine translation of rare words with subword units. arXiv preprint arXiv:1508.07909 (2015)
23. Siivola, V., Hirsimäki, T., Creutz, M., Kurimo, M.: Unlimited vocabulary speech recognition based on morphs discovered in an unsupervised manner. In: Proc. Eurospeech. vol. 3, pp. 2293–2296 (2003)
24. Yuret, D., Biçici, E.: Modeling morphologically rich languages using split words and unstructured dependencies. In: Proceedings of the ACL-IJCNLP 2009 conference short papers. pp. 345–348 (2009)

Estimation of Hyperspectral Images Bands Similarity Using Textural Properties

M.I. Khotilin[1][0000-0001-7987-7216], R.A. Paringer[1,2][0000-0003-0560-8608], A.V. Kupriyanov[1,2][0000-0002-0436-4392], D.V. Kirsh[1,2][0000-0002-3917-5444], D.G. Asatryan[3,4][0000-0000-0000-0000], M.E. Haroutunian[3][0000-0002-9262-4173], A.V. Mukhin[1][0000-0003-1738-3098]

[1] Samara University, 34, Moskovskoye shosse, Samara, 443086, Russia
[2] Image Processing Systems Institute of the Russian Academy of Sciences – Branch of the FSRS "Crystallography and Photonics" RAS, Samara, Russia
[3] Institute for informatics and automation problems of NAS Armenia, P. Sevaki Str. 1, Yerevan, 0014, Armenia
[4] Russian-Armenian university, Hovsep Emin Str. 123, Yerevan, 0052, Armenia
khotilin.mi@ssau.ru rusparinger@gmail.com

Abstract. In this paper, the technology for generating a set of effective features based on the analysis of the textural properties of given image classes using discriminant analysis is used to analyze the properties of hyperspectral images. A set of hyperspectral images of plant leaves is used, with a resolution of 64×64 pixels and the number of spectral channels equal to 237 (wavelengths from 436 nm to 965 nm). It was shown that the definition of not just features, but their properties and the application of knowledge about the interconnetedness of different types of features and hyperspectral data of different wavelengths can be used to pre-process hyperspectral images in order to reduce the computational complexity of algorithms for hyperspectral images analysis. In result were estimated that textural features mostly useful for analysis task that used wavelength before 450 nm, run-length features useful for more than 900 nm.

Keywords: Hyperspectral Images, Classification, Feature Selection, Textural Features, Feature Aggregation.

1 Introduction

Hyperspectral images are three-dimensional data arrays that include spatial information about an object, supplemented with spectral information for each spatial coordinate [1]. Currently, processing and analysis of hyperspectral images are popular research topics in the field of image processing and computer vision. Within the framework of this article, the technology of constructing and applying informative features of a hyperspectral image for the problem of image classification is considered.

2 Technology for the Research of the Informative Features for the Analysis of Hyperspectral Images

In the course of this work, there was used the technology for the formation of a set of effective features based on the analysis of the textural properties of these classes of images using discriminant analysis proposed by the author in [9].

For the study and processing of image layers, it was decided to use the well-proven MaZda software [4], which makes it possible to calculate various texture groups of features. As a result of the work of this software product, we obtain a set of features and their values, which will be used in the future.

The next step is the aggregation of the obtained features and values for each of the layers. This procedure allows to consider the distribution of the values of each of the features in images of different classes.

Next, features are selected for each layer by one of three methods: pairwise, general, one from the rest.

After analysis of the frequency of occurrence of features on different layers, the threshold filtering was used, which filters out features that are less common than the threshold value.

Combining features by type, as, for example, in [8], we can conclude which features are significant for distinguishing given classes of images and how the selected features depend on the wavelength of the studied hyperspectral images.

3 Practical Research and Implementation

In accordance with the presented technology, at the first stage of selecting the initial data, sets of hyperspectral images 64 × 64 pixels in size, with the number of spectral channels equal to 237, wavelengths from 436 nm to 965 nm, were considered. These images were pictures of the leaves of various plants, such as: tomato, pepper, cabbage, carrot and others.

The next step is to research the previously obtained images and extract the features characterizing them using the MaZda software. An array of images-layers is used as input data, and as a result of processing the input data, the software presents a report containing various characteristics and features [8]. Combining all the report data, we get a list of features that characterize images of one or another class.

Next comes the stage of selecting informative features for each of the layers. In this work, the general, pairwise and "one from the rest" methods were used for selecting informative features. With the general method of feature selection, the feature space is composed of the best features according to the value of the discriminant analysis separability criterion calculated for all classes of objects under study. With the pairwise feature selection method, the essence is to select the features that have the highest value of the separability criterion when distinguishing each pair of classes and further combining them into a single set. All the features selected for different layers were counted and ranked based on the frequency of occurrence in the number of informative ones for distinguishing the given classes on different layers of hyper-spectral data.

Next, we analyze the frequency of occurrence of features on different layers. The results are presented in table 1.

Table 1. Selected features and number of references to them.

General selection		Pairwise selection		One from the rest	
Number of references	Feature name	Number of references	Feature Name	Number of references	Feature name
237	Perc.01%	237	Perc.01%	236	Perc.01%
237	Perc.10%	232	Perc.10%	236	Perc.10%
237	Perc.50%	231	Perc.90%	229	Perc.50%
233	Teta3	222	Perc.50%	188	Teta3
206	Perc.90%	218	S(4,0)Correlat	150	S(4,0)Correlat
199	Teta1	216	S(5,0)Correlat	146	Vertl_ShrtREmp
167	Skewness	215	Teta3	145	S(5,0)Correlat
162	Vertl_ShrtREmp	213	S(3,0)Correlat	131	S(3,0)Correlat
157	Perc.99%	202	Perc.99%	128	Vertl_Fraction
149	Teta2	189	Teta1	112	Teta1
135	Horzl_Fraction	177	S(2,0)Correlat	96	Horzl_Fraction
133	S(2,0)Correlat	159	Vertl_ShrtREmp	77	135dr_Fraction
129	S(3,0)Correlat	139	S(0,1)DifEntrp	67	45dgr_Fraction
129	Vertl_Fraction	136	S(0,5)InvDfMom	65	Teta2
102	135dr_Fraction	131	Vertl_Fraction	62	WavEnHH_s-1
102	GrKurtosis	121	S(0,4)InvDfMom	60	S(2,0)Correlat
100	S(4,0)Correlat	120	S(3,3)Correlat	56	S(5,0)InvDfMom
99	WavEnHH_s-1	116	S(0,2)DifEntrp	54	S(4,0)InvDfMom
97	S(0,1)DifEntrp	114	S(0,3)DifEntrp	51	GrKurtosis
95	45dgr_Fraction	111	S(2,2)Correlat	48	S(0,2)DifEntrp
89	S(1,0)DifEntrp	103	S(4,4)Correlat	46	S(0,1)DifEntrp
87	S(0,2)DifEntrp	101	S(1,0)DifEntrp	40	S(0,5)InvDfMom

For further work, features that occur more often than on 42% of the layers (100 pieces) of the research images were selected. Since different types of features reveal different properties of images, the informative features selected in the previous stage were combined into sets, depending on the type: histogram, texture, path lengths, autoregressive, gradient. Thus, an analysis of the properties revealed by one or another set of features, depending on the layer of hyperspectral data used was made. The results are presented in figures 1-3.

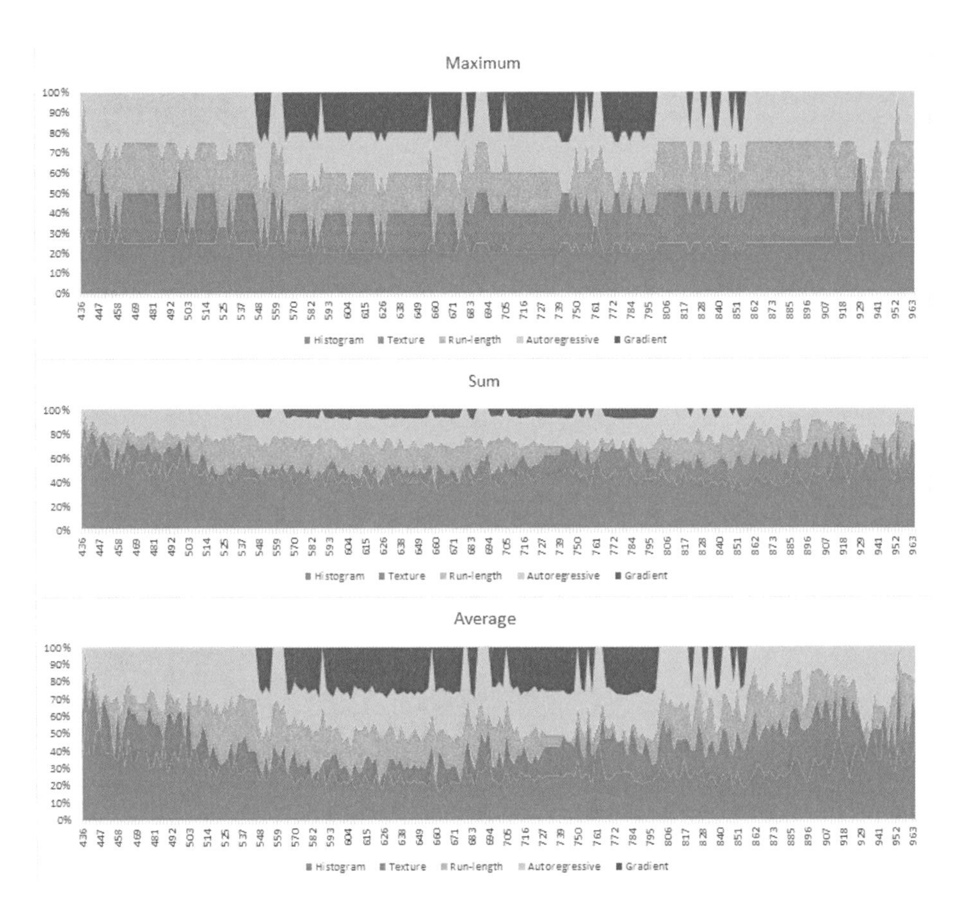

Fig. 1. Plots of maximum, total and average for general selection.

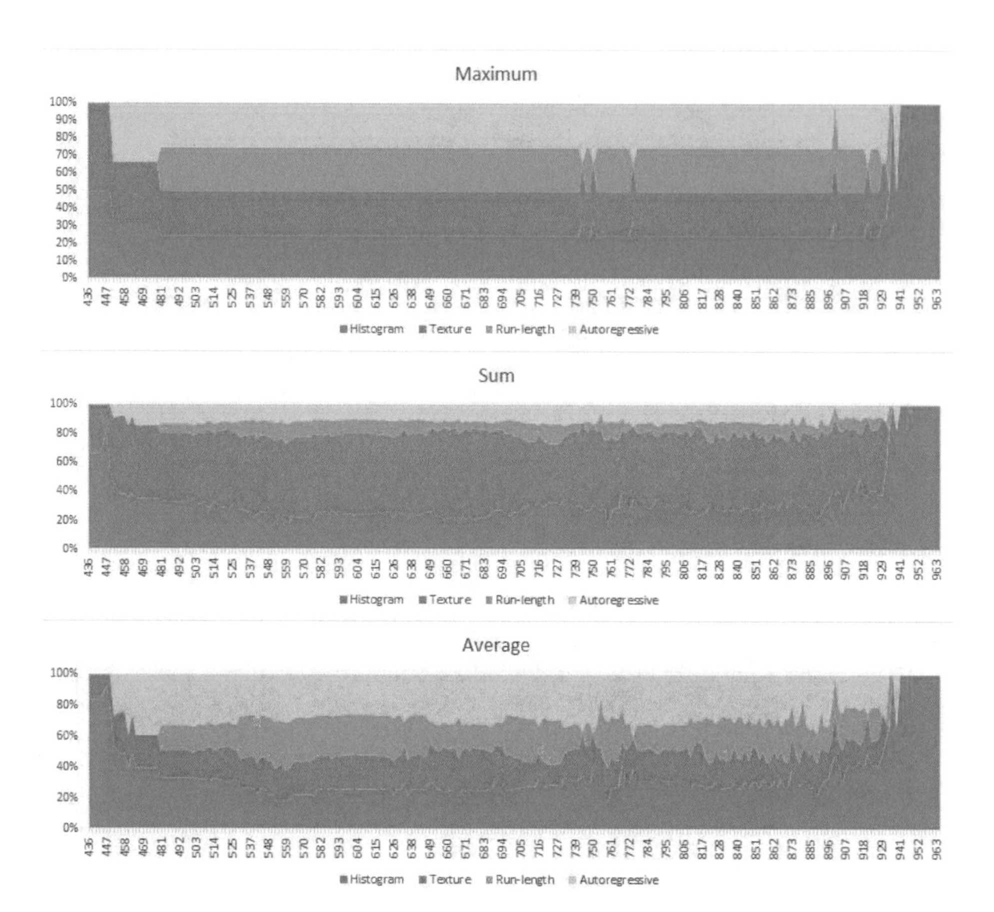

Fig. 2. Plots of maximum, total and average for pairwise selection.

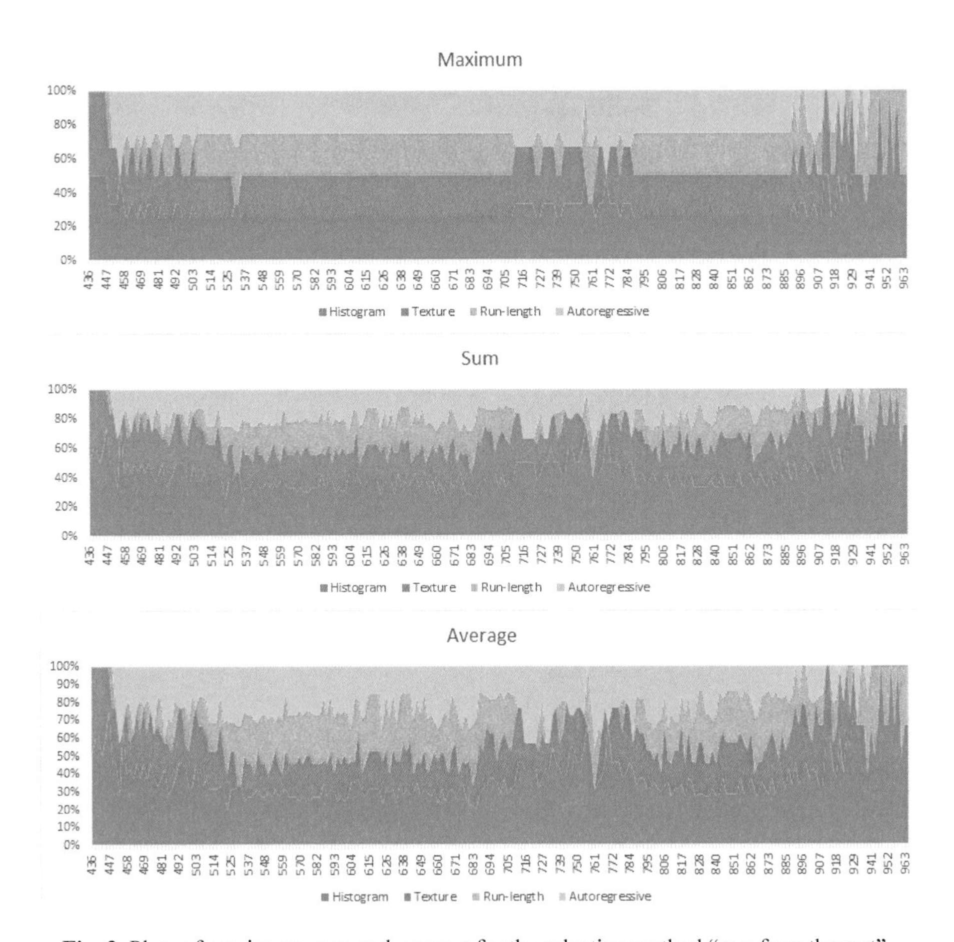

Fig. 3. Plots of maximum, sum and average for the selection method "one from the rest".

When considering the graphs, it was noticed that the method of pairwise feature selection selects predominantly textural features.

As can be seen from the data above, textural features perform better at wavelengths corresponding to the blue range of visible light and near infrared radiation. This situation may indicate that, due to the fact that the initial data in this work are a photo of a plants, the differences between the classes in the red and green light ranges may not be clearly defined. The blue range, on the other hand, allows to observe differences in texture, and therefore textural features in the blue range are applicable better than others. When considering the infrared range, it can be concluded that due to the disappearance of textural features when considering certain wavelengths, as the wavelength increases, noise components appear that contribute to the disappearance of textural features, i.e. the texture becomes indistinguishable viewed.

4 Conclusion

The results obtained in this work can be used to develop tools for the intellectual analysis of hyperspectral images of various areas of human life and activity. For example, in agriculture, examining images taken from fields occupied by certain crops, one can find those that are weedy or that are different from those growing originally. Also, when compiling forest maps, it is possible to determine the composition of forests by hyperspectral images of their surface. Determination of not just features, but their properties and application of knowledge about the interconnectedness of different types of features and hyperspectral data of different wavelengths can be used to pre-process hyperspectral images in order to reduce the computational complexity of the developed algorithms. According to obtained results textural features are mostly useful for wavelength before 450 nm, run-length features useful for more than 900 nm and useless for 700-770 nm.

Further development of this work is aimed at researching and creating a complex technology that allows clustering and classifying hyperspectral image objects and significantly reducing time and computational resources. Experimental results can be used to develop algorithms to create high-precision analysis tools using portable wearable devices.

5 Acknowledgements

This work was supported by the Russian Foundation for Basic Research and RA Science Committee in the frames of the joint research project RFBR 20-51-05008 Arm_a and SCS 20RF-144 accordingly.

References

1. Zimichev, E.A., Kazanskiy, N.L., Serafimovich, P.G.: Spectral-spatial classification with k-means++ particional clustering. Computer Optics 38(2), 281-286(2014)
2. Sergeev, V.V.: Application of pattern recognition methodology in digital image processing tasks. Autometry 2, 63-76 (1998)
3. Kazanskiy, N.L., Kharitonov, S.I., Khonina, S.N., Volotovskiy, S.G., Strelkov, Yu.S.: Simulation of hyperspectrometer on spectral linear variable filters. Computer Optics 38(2), 256-270 (2014)
4. Strzelecki, M.: A software tool for automatic classification and segmentation of 2D/3D medical images. Nuclear Instruments and Methods in Physics Research Section A: Accelerators, Spectrometers, Detectors and Associated Equipment 702, 137-140 (2013). doi:10.1016/j.nima.2012.09.006
5. Khotilin, M., Kravtsova, N., Rytsarev I., Kupriyanov, A.: Classification of objects of natural hyperspectral images. In: 2020 International Conference on Information Technology and Nanotechnology (ITNT), pp. 1-3. IEEE, New York (2020). doi: 10.1109/ITNT49337.2020.9253254
6. Goncharova, E.F.: Greedy algorithms of feature selection for multiclass image classifica-

tion. In: CEUR Workshop Proceedings (IPERS-ITNT 2018-Proceedings of the International Conference on Information Technology and Nanotechnology - Session: Image Processing and Earth Remote Sensing), vol. 2210, pp. 38-46. Samara University (2018)

7. Khotilin, M.: The technology of constructing an informative feature of a natural hyperspectral image area for the classification problem. International Conference on Information Technology and Nanotechnology (ITNT), pp. 1-4. Samara University (2021). doi: 10.1109/ITNT52450.2021.9649178
8. Fruehwald-Pallamar, J.: Texture-Based Analysis of 100 MR Examinations of Head and Neck Tumors - Is It Possible to Discriminate Between Benign and Malignant Masses in a Multicenter Trial? RoFo: Fortschritte auf dem Gebiete der Röntgenstrahlen und der Nuklearmedizin 188, 195–202 (2016) doi: 10.1055/s-0041-106066
9. Shirokanev, A., Ilyasova, N., Paringer, R.: A smart feature selection technique for segmentation of fundus images. Procedia Engineering 201, 736-745 (2018)
10. Ilyasova, N., Paringer, R.A, Kupriyanov, A.V.: Regions of interest in a fundus image selection technique using the discriminative analysis methods. In: ICCVG 2016, LNCS, vol. 9972, pp. 408-417. Springer, Heidelberg (2016). https://doi.org/10.1007/978-3-319-46418-3_36
11. Zhang, J., Geng, W., Liang, X., Li, J., Zhuo, L., Zhou, Q.: Hyperspectral remote sensing image retrieval system using spectral and texture features. Appl. Opt. 56(16), 4785-4796 (2017). https://doi.org/10.1364/AO.56.004785
12. Zhang, X., Zhang, K., Sun, Y., Zhao, Y., Zhuang, H., Ban, W., Chen, Y., Fu, E., Chen, S., Liu J, Hao, Y.: Combining Spectral and Texture Features of UAS-Based Multispectral Images for Maize Leaf Area Index Estimation. Remote Sensing 14(2), 331-352 (2022). https://doi.org/10.3390/rs14020331

Reconstruction of Convex Bodies by Tomographic Methods*

V. K. Ohanyan[0000-0001-7029-2385]

American University of Armenia and Yerevan State University
victo@aua.am, victoohanyan@ysu.am

Abstract. The present report contains a review of the main results of Yerevan research group in tomography of bounded convex bodies. Let $\mathbf{R}^n$ $(n \geq 2)$ be the n-dimensional Euclidean space, $\mathbf{D} \subset \mathbf{R}^n$ be a bounded convex body. Random k-flats in $\mathbf{R}^n$, $1 \leq k \leq n-1$ generate cross sections of random size in convex body $\mathbf{D}$. As $\mathbf{D}$ is a convex body, then obviously intersections of k-flats with $\mathbf{D}$ are always connected subsets of $\mathbf{R}^n$ for every $k \in \{1, ..., n-1\}$. The determination of the distribution of size of cross sections has a long tradition of application to collections of bounded convex bodies forming structures in metal and crystallography. However, calculations of geometrical characteristics of random cross sections is often a difficult task. In a special case $k = 1$ we call the corresponding distribution function as the chord length distribution function. For $n = 2$ the list of known results was expanded after 2005 when N. G. Aharonyan and V. K. Ohanyan obtained the explicit formula of the chord length distribution function for a regular polygon. A computer program is created which gives values of a chord length distribution function in the case of a regular n-gon for every natural $n \geq 3$. These all problems have applications in Medicine Tomography.

Keywords: Stochastic geometry · Covariogram · Chord length distribution · Tomography.

1 The Main Results

Complicated geometrical patterns occur in many areas of science. Their analysis requires creation of mathematical models and development of special mathematical tools. The corresponding area of mathematical research is called Stochastic Geometry (see [7]). Among more popular applications are Stereology and Tomography (see [8], [7] and [2]).

Reconstruction of the body over its cross section is one of the main tasks of geometric tomography, a term introduced by R. Gardner in [7]. If $D \subset \mathbb{R}^n$ ($\mathbb{R}^n$ is n-dimensional Euclidean space) is a compact convex body, it is possible to intersect it by a random k-flats $(1 < k \leq n-1)$. If the body D is intersected

* The investigation is done with partial support by the Mathematical Studies Center at Yerevan State University.

by k-flats, then a k-dimensional section contains some information on D. The natural question arises whether it is possible to reconstruct D, if we have a subclass of k-dimensional cross-sections. Reconstruction of convex bodies using random sections makes it possible to simplify the calculation, since the estimates of probability characteristics can be obtained using the methods of mathematical statistics. Quantities characterizing random sections of the body D (such as the random chord length, the chord length in the random direction, the surface area of a random section of the body formed by the intersection of a random k-plane, and others) carry some information on D and if there is a connection between the geometrical characteristics of D and probabilistic characteristics of random cross-sections, then by a sample of results of experiments we can estimate the geometric characteristics of the body D. The question of the existence of a bijection between bounded convex bodies D and distribution functions of the chord length $F_D(x)$ was made by the famous German mathematician Wilhelm Blaschke. Further mathematics considered subclasses of bounded convex bodies for which the chord length distribution function reconstructed non-congruent elements of a subclass. Although the chord length distribution function $F_D(x)$ does not reconstruct the compact convex body, yet it contains information about the volume, surface area and other characteristics of the body.

Denote by S^{n-1} the $(n-1)$-dimensional sphere of radius 1 centered at the origin. Let $\Pi r_{u^\perp} D$ the orthogonal projection of D on the hyperplane $u^\perp (u^\perp$ is the hyperplane with normal u and passing through the origin).

Definition 1 *The covariogram of a convex body D is defined by*

$$C_D(x) := V_n(D \cap \{D + x\}), \quad x \in \mathbb{R}^n, \tag{1}$$

where $D+x = \{P+x : P \in D\}$ and $V_n(\cdot)$ is the n-dimensional Lebesgue measure.

$C_D(x)$ is invariant with respect to translations and reflections. The same problem was posed independently in the context of probability theory.

Whether the distribution of the difference $\eta_1 - \eta_2$ of two independent random variables η_1 and η_2 that are uniformly distributed over D determines D up to translation and reflection?

It follows from the following formulae:

$$C_D(x) = \int_{\mathbb{R}^n} I_D(y)\, I_{D+x}(y)\, dy = \int_{\mathbb{R}^n} I_D(y)\, I_D(y-x)\, dy,$$

where $I_D(\cdot)$ – the indicator of the set D, while

$$f_{\eta_1-\eta_2}(x) = \frac{1}{V_n^2(D)} \int_{\mathbb{R}^n} I_D(y)\, I_D(y-x)\, dy = \frac{C_D(x)}{V_n^2(D)}, \tag{2}$$

where $f_{\eta_1-\eta_2}(x)$- the density function of the distribution of the difference $\eta_1 - \eta_2$.

Note, that the density function $f(x)$ of each random variable η_i, $i = 1, 2$ equals:

$$f(x) = \frac{I_D(x)}{V_n(D)}. \tag{3}$$

The problem also arises in the Fourier analysis:

Determining a function γ, with compact support in $\mathbb{R}^n$ from the modulus of its Fourier transform (characteristic function, if γ is a density function of some random variable)? It follows from a remark, that the characteristic function of a difference of two independent and identically distributed random variables η_1 and η_2 has the following form:

$$|\varphi(t)|^2$$

where $\varphi(t)$ is the characteristic function of the random variable η_i, $i = 1, 2$, and as the function γ we have to take normalized indicator function of form (3) (see [17]).

Therefore, the last problem reduces to the covariogram problem, because taking the characteristic function of the right- and left-hand side of formula (2) we obtain:

$$\varphi_{C_D(x)}(t) = |\varphi(t)|^2$$

where $\varphi_{C_D(x)}(t)$ is the characteristic function for $\frac{C_D(x)}{V^2(D)}$, while $\varphi(t)$ is the characteristic function of the random variable (3).

G. Matheron proved, that for any $t > 0$ and $\varphi \in S^{n-1}$

$$\frac{\partial C_D(t\varphi)}{\partial t} = -V_{n-1}(\{y \in \varphi^{\perp} : \; L_1(D \cap (l_\varphi + y)) \geq t\}), \tag{4}$$

where $l_\varphi + y$ is the line, parallel to the direction φ through the point y, while $\varphi^{\perp}$ denotes the orthogonal complement of φ, that is, the hyperplane in $\mathbb{R}^n$ with normal direction $\varphi \in S^{n-1}$.

Let $\mathbf{G}$ be the space of all lines in the Euclidean plane $\mathbb{R}^2$, $g \in \mathbf{G}$, and (p, φ) are the polar coordinates of the foot of the perpendicular to g from the origin; $p \geq 0$, $\varphi \in S^1$. For a closed bounded convex domain $D \subset \mathbb{R}^2$ we denote by $S_D(\varphi)$ the support function in direction $\varphi \in S^1$ defined by

$$S_D(\varphi) = \max\{p \in \mathbb{R}^+ : \; g(p, \varphi) \cap D \neq \emptyset\},$$

where $\mathbb{R}^+$ is the set of nonnegative real numbers.

For a bounded convex domain $D \subset \mathbb{R}^2$ we denote by $b_D(\varphi)$ the breadth function in direction $\varphi \in S^1$, that is, the distance between two support lines to the boundary of D that are perpendicular to φ. We have

$$b_D(\varphi) := S_D(\varphi) + S_D(\varphi + \pi).$$

Note that $b_D(\varphi)$ is a periodic function with period π.

A random line which is parallel to u and intersects D has an intersection point (denote by x) with $\Pi r_{u^\perp} D$. Assuming that the intersection point x is uniformly distributed over the convex body $\Pi r_{u^\perp} D$ we can define the following distribution function:

Definition 2 *The function*

$$F_D(u, t) = \frac{V_{n-1}\{x \in \Pi r_{u^\perp} D : V_1(g(u, x) \cap D) < t)\}}{b_D(u)} \tag{5}$$

is called orientation-dependent chord length distribution function of D in direction u at point $t \in R^1$, where $g(u,x)$ is the line which is parallel to u and intersects $\Pi r_{u^\perp} D$ at point x and $b_D(u) = V_{n-1}(\Pi r_{u^\perp} D)$.

Matheron formulated a hypothesis that there exists a one-to-one correspondence between $F_D(u,t)$ and bounded convex bodies. Matheron's hypothesis received a positive solution for any D in the planar case. In the case of finite-dimensional spaces with $n > 3$ Matheron's hypothesis has received a negative answer. Very little is known regarding the covariogram problem when the space dimension is greater than 2. In the case of 3-dimensional space the problem is open. Nevertheless, for the case of bounded convex polyhedron for $n = 3$ Matheron's hypothesis received a positive answer (see [5]). It is known that centrally symmetric convex bodies in any dimension, are determined by their covariogram up to translations. We note that in the case where there is Matheron's hypothesis, the authors prove the existence and uniqueness of D, but do not construct the corresponding unique D. Hence it is important the works of Yerevan group of mathematicians which gives explicit expressions both the covariogram and the distribution functions $F_D(u,t)$ and $F_D(t)$ for a broad class of convex bodies as in the plane as in the 3-dimensional space (see [12] and [13]). A practical application these results in crystallography can be found in [4] and [12].

The natural question arises whether it is possible to reconstruct the body having values of $F_D(u,x)$ only for a finite set of directions. This question received negative answer, because it is possible to construct two non-congruent triangles that have the same chord length distribution function for a fixed set of m directions, where m is a natural (see [15]). The question arises whether it is possible to find a subclass of convex bodies, where it is possible to reconstruct a body from the values of $F_D(u,x)$ for a finite set of directions.

Till recently explicit expressions for the chord length distribution functions have been known in the case when $\mathbb{D}$ is a disc, a regular triangle and a rectangle (see [7] and [1]). These results have been obtained using the definition of chord length distribution function for a domain $\mathbb{D}$.

In the recent years our group has obtained important results in this direction. We have obtained explicit expression of the chord length distribution function for any regular polygon (see [8]). In the particular cases of a regular triangle, a square, a regular pentagon and a regular hexagon our formula for the chord length distribution function coincides with formulas available in the literature (see [1] — [3], [10], [16] and [6]) for $n = 3$, 4, 5 and 6 correspondingly.

In the last years the notion of orientation-dependent chord length distribution function have been introduced. In the paper [14] an explicit formula for orientation-dependent chord length distribution function for any bounded convex domains $\mathbb{D}$ have been obtained. These questions are connected with Covariogram Problem: Does the covariogram determine a convex body, among all convex bodies, up to translations and reflections? G. Matheron conjectured a positive answer for this problem, [1] and [9]. In fact, the covariogram problem is equivalent to the problem of determining a convex body from all its orientation-dependent chord length distributions (see [5] and [11]). All these problems are the problems of geometric tomography ([7]), since orientation-dependent chord length distribution function is the probability that parallel X-ray in direction ϕ less than or equal to y.

References

1. Schneider, R., Weil, W.: Stochastic and Integral Geometry. Springer-Verlag Berlin Heidelberg (2008).
2. Ohanyan, V. K.: Combinatorial principles in Stochastic Geometry: A Review. Journal of Contemporary Mathematical Analysis **43**(1), 44 – 60 (2008).
3. Aharonyan, N. G., Ohanyan, V. K.: Chord length distribution functions for polygons. Journal of Contemporary Mathematical Analysis **40**(4), 43 – 56 (2005).
4. Gille, W., Aharonyan, N. G., Harutyunyan, H. S.: Chord length distribution of pentagonal and hexagonal rods: relation to small-angle scattering. Journal of Applied Cristallography **42**, 326 – 328 (2009).
5. Bianchi, G., Averkov, G.: Confirmation of Matheron's Conjecture on the covariogram of a planar convex body. Journal of the European Mathematical Society **11**, 1187 – 1202 (2009).
6. Gasparyan, A., Ohanyan, V. K.: Covariogram of a parallelogram. Journal of Contemporary Mathematical Analysis **49**(4), 194 – 206 (2014).
7. Gardner, R. J.: Geometric Tomography. 2nd edn, Cambridge University Press, Cambridge, UK, New York (2006).
8. Harutyunyan, H., Ohanyan, V.K.: Chord length distribution function for regular polygons. Advances in Applied Probability **41**, 358 – 366 (2009).
9. Santalo, L. A.: Integral Geometry and Geometric Probability. Addision-Wesley, Reading (2004).
10. Burgstaller, B., Pillichshammer, F.: The average distance between two points. Bull. Aust. Math. Soc. **80**, 353 – 359 (2009).
11. Ohanyan, V. K., Martirosyan, D. M.: Orientation-dependent chord length distribution function for right prisms with rectangular or right trapezoidal bases. Journal of Contemporary Mathematical Analysis **55**(6), 68 – 82 (2020).
12. Aharonyan, N. G., Ohanyan, V. K.: Calculation of geometric probabilities using Covariogram of convex bodies. Journal of Contemporary Mathematical Analysis **53**(2), 110 – 120 (2018).
13. Harutyunyan, H. S., Ohanyan, V. K.: Covariogram of a cylinder. Journal of Contemporary Mathematical Analysis **49**(6), 366 – 375 (2014).
14. Aharonyan, N. G.: Generalized Pleijel identity. Journal of Contemporary Mathematical Analysis **43**(5), 3 – 14 (2008).
15. Gasparyan, A., Ohanyan, V. K.: Recognition of triangles by covariogram. Journal of Contemporary Mathematical Analysis **48**(3), 110 – 122 (2013).
16. Aharonyan, N. G., Khalatyan, V.: Distribution of the distance between two random points in a body from R^n. Journal of Contemporary Mathematical Analysis, **55**(6), 329 – 334 (2020).
17. Bianchi, G., Averkov, G.: Retriving convex bodies from restricted covariogram functions. Adv. Appl. Prob. (SGSA) **39**, 613 – 629 (2007).

Validation of Risk Assessment Models for Breast and Ovarian Cancer–Related Gene Variants

Wolfram Luther [0000-0002-1245-7628]
University of Duisburg-Essen, Germany
wolfram.luther@uni-due.de

Abstract. Quality criteria and metrics are part of a coherent framework for validating results in medical studies and for comparing mathematical and statistical risk models for the occurrence of certain diseases in a cohort under consideration. Metrics, such as the Dempster-Shafer basic probability assignment or bar charts, for the occurrence of hereditary pathogenic variants in BRCA genes at age intervals for the first diagnosis of breast or ovarian cancer in an individual or their relatives are developed using patient data from clinical studies and can be used to validate mathematical and statistical models for the occurrence and impact of hereditary pathogenic variants in a cohort and its subgroups. In turn, model-based metrics can be used to validate consistent assignment of the cohort or its constituent parts from a study of subjects at risk of hereditary breast and ovarian cancer syndromes to one of several mutation risk classes. Using the two general metrics introduced, we reproduced mean values for the occurrence of deleterious BRCAx gene variants for sporadic, mixed and high-risk breast cancer cohorts of varying ethnicities and their subgroups with specific disease patterns from five large international studies.

Keywords: Pathogenic BRCA Variants, Family Cancer History, Quality Metrics Assessment, Genetic Testing.

1 Introduction

Cancer occurs from pathogenic genetic variants (also referred to as *mutations*) that involve changes in the order of the base pairs. Women with a BRCAx gene mutation have a six times higher lifetime risk (70%) of developing breast cancer (BC) than noncarriers. For women in the general population, ovarian cancer (OC) is roughly six times less common in women than BC, while the cumulative risk to age 80 is 44% for BRCA1 carriers and 17% for BRCA2 carriers [26]. Somatic variants are the most common cause of cancer, occurring from damage to genes in an individual cell during a person's life and are responsible for BC of various subtypes, such as luminal (estrogen receptor, HER2 and basal-like cancer, and triple negative TNBC), at any age. We differentiate between postmenopausal BC (BC_{51+}) at age 51 or older and premenopausal (BC_{50-}) with first age of onset (fao) of 50 years or younger as well as between rare ovarian cancers OC and OC_{50-} diagnosed at earlier ages with somatic TP53 mutations and germline BRCA1/2 variants [21, 35].

Germline variants accounting for only about 5%–10% of all cancers occur in reproductive cells and are passed directly from a parent to a child. Family histories (FH) and individuals suggestive of germline pathogenic variants include earlier onset cancers of various types, subtypes and disease patterns (TNBC, forms of bilateral/ multifocal BBC, male breast cancer MBC, breast and ovarian cancer BCOC, BC and OC in one person (OBC), colon, pancreatic or prostate cancer etc.). These diseases are the result of autosomal-dominant inheritance from the paternal or maternal line to both sexes over several generations resulting in increased risk of disease at an earlier age, e.g., BC_{40-}, which further increases in ethnic groups with founder effects, such as Ashkenazi Jews (AJ). The individual risk for the occurrence of a pathogenic gene variant (PV) depends, among other things, on the number of cancers in the FH, the age of their first occurrence, and degree of kinship with the index patient (IP) with the earliest age of BC diagnosis. In the case of germline mutations, the risk to a (healthy) family member in the respective adjacent generation is halved.[1]

Jervis et al. [23] state that cancer occurrence in FH is one of the most important risk factors for developing epithelial ovarian cancer (EOC). Average values of OC familial relative risks (FRRs) depend highly on the probe's BRCA1 and BRCA2 mutation status: BRCA- 2.24%, BRCA1+ 20.97%, and BRCA2+ 9.57%.

Peshkin and Buys [9, 34] give an up-to-date overview on BRCAx and multigene panel testing and management of individuals at risk of hereditary breast and ovarian cancer syndromes (abbreviated HBOC), including risk-reducing prevention strategies. They advocate intensified screening to increase the number of individuals identified as carrying a pathogenic variant in the cohort with a special focus the ethnic composition, first age of diagnosis and TNBC status. The frequency of pathogenic variants is visualized with meaningful bar charts by age in five-year increments between 20 and 90 years in all individuals with breast cancer.

Modern genetic tests reliably identify BRCA1/2 mutations, but are not necessarily helpful for everyone. Due to the relative rarity of hereditary mutations and the consequences of a positive test result for the affected person and his/her family members, various offers of counseling by specialists and relevant institutions should be made in advance. In evidence reviews by Nelson, Owens et al. [28, 30], it is stated that in some situations individual testing can also be replaced by a strategy of population-based screening for "founder mutations that detect more BRCA1/2 mutation carriers than testing persons who met family history criteria."

In medical practice, there exist fairly accurate numerical tools and publicly available mathematical statistical models/algorithms predicting age specific BC or OC probability (or gene mutation status) based on such factors as the patient's FH, breast tumor molecular marker information and frequencies of occurrence in the population to which the cohort is to be assigned. BRCAPRO[2], BOADICEA, now CANRISK[3], Claus [11], and PENN II[4] are important mathematical risk models with associated software for computing the probability of deleterious gene mutation based on Mendelian genetics and the Bayes theorem [6, 11, 14]. The PENN II risk model is an internet survey form with ten concrete questions about the FH of cancer on one side of the family that predicts the likelihood of BRCA1/2 mutation (lBRCAm) of the patient/proband or the closest first/second-degree relative with cancer.

Widely used questionnaire-type risk assessment tools, such as the Ontario Family History Assessment Tool (FHAT), Manchester Scoring System (MSS), or Pedigree Assessment Tool (PAT), evaluate cancers in the proband's FH with a point table for

[1] https://voice.ons.org/news-and-views/germline-and-somatic-mutations-what-is-the-difference
[2] https://projects.iq.harvard.edu/bayesmendel/brcapro
[3] https://www.canrisk.org/guide
[4] https://pennmodel2.pmacs.upenn.edu/penn2/

BRCA1/2 depending on degree of kinship and fao or simply by number of occurrences (Referral Screening Tool, RST). They recommend genetic testing when a certain threshold is reached or exceeded, signifying inclusion in the class of elevated lBRCAm [2, 24, 28, 30, 31, 32, 37].

This process is highly affected by uncertainty as information about FH and age is often incomplete or inaccurate at first diagnosis. To include epistemic uncertainty due to missing fao information, the extended referral screening tool (ERST) combines the binary decision structure of an RST with the features of the FHAT, the MSS and Frank et al. [16] in an accumulated interval risk function. The ERST also produces a referral interval vector U for eight decision paths to help assign participants to low, moderate and high-risk categories [3]. At the same time, the reliability and adequacy of genetic counseling and testing is especially important since they directly influence humans and allow the specification of the average or interval-based mutation risk of an IP and his/her relatives or their assignment to a risk class (average/slow, moderate/elevated, high) and decisions about the patient's further care and treatment [31].

Therefore, large studies on the frequency of occurrence of pathogenic variants with a cohort of IPs selected according to standardized criteria (mostly younger patients with one or more cancers and their families), such as those conducted by Fischer, Frank, Kast, Hall, and Okano [7, 15, 16, 19, 25, 29], are of great importance because they provide accurate data on the number of BRCA1+, BRCA2+, and BRCA- cases. Furthermore, they enable a comparative evaluation of the risk models and counseling tools according to specified quality criteria and metrics with respect to the predictions and assignment of disease patterns to risk classes and thus help to avoid harms associated with erroneous risk assessment, excessive genetic counseling and testing, or unnecessary interventions. On behalf of the US Preventive Services Task Force and national HBOC consortia, Owens, Evans, and Fischer [13, 14, 30] present recommendations on risk assessment models, genetic counseling, and genetic testing for BRCA-related cancer, evaluate possible interventions for affected patients and families, and assess their impact magnitude for the possible scenarios according to risk classes. Studies examining individuals with their family history and BC together with OC diseases for pathogenic gene variants raise a principled problem. Since the proportion of OC diseases is on average only 17% of BC diseases—that is, they occur much less frequently—the estimates for the influence of OC diseases on the lBRCAm value (i.e., the percentage of deleterious mutations) will be inaccurate and underestimated.

This has been shown by studies that chose a priori cohorts for testing in which all index patients had OC, as in Gupta [18], or excluded families with only one OC case [25], resulting in a lBRCAm close to 25%. Predictions by experts, which they obtain from databases on the gene mutations and their carriers, point in the same direction.

In this work, we present quality criteria and their metrics to estimate the risk of an index patient and his/her relatives for the occurrence of pathogenic BRCAx variants, taking into account the origin of each individual and his/her family members, specific selection criteria, disease constellations, and initial diagnosis data for the entire cohort and its typical subgroups. Section 1 introduces terminology and reviews risk prediction models, genetic counseling and testing. Section 2 develops quality criteria and metrics, data reliability assessment, and risk modeling and computing for germline mutation in BRCA genes. Section 3 deals with data management and quality assessment of risk studies determining the frequency of occurrence of PV in BRCA genes. In it, we introduce probability metrics for the individual and familial incidence of pathogenic variants depending on disease types and subtypes, their combinations and age of first diagnosis in the form of bar charts with time period sections and Dempster-Shafer basic probability assignments (BPA). The metrics provide estimates of the proportions of carriers and noncarriers, as well as the lBRCAm in the cohort and in its subgroups with a sample of BC, OC, and BCOC diseases. These metrics allow reciprocal statements on their suitability and on quality criteria such as performance, accuracy, consistency for genetic counseling, risk models and comparative surveys. Reliable metrics achieve accurate and consistent results, while the quality of risk study results can be determined with these metrics. Section 4 gives our conclusions and highlights further work.

2 Requirements for Concise Validation Management

2.1 Quality Criteria and Metrics in Genetic Counseling and Testing

Validation and verification (V&V) assessment describes options along with reliability domains, quality criteria (QC) and quality metrics (QM). One of the fundamental domains of assessment is resource and data management with QC accuracy, consistency, completeness, integrity and other technical aspects, such as availability and readability [10]; a second domain is devoted to mathematical modeling and computing. Here, the QC refer to accuracy, performance (i.e., efficiency and effectivity) and adequacy. Further domains could be clinical genetic testing [27], visual analytics [4], risk perception, decision-making and knowledge co-creation [7]. Cross-sectional issues are the consideration of different forms of uncertainty and the realization of interdisciplinary collaboration.

As is mentioned in [3, 5], there are many studies and meta-studies that review/ evaluate genetic counseling tools and risk assessment models based on ground truth data as is highlighted in [1, 3, 7, 20, 28, 30, 32] among a multitude of other publications. However, they focus mainly on aleatoric uncertainty and consider epistemic uncertainty indirectly through comparisons. Examples of QM used in lBRCAm risk assessment include sensitivity, specificity, discriminatory accuracy and receiver operating characteristic analysis [27, 31].

Aleatoric uncertainty is usually measured through a confidence interval around the mean result. In [3] intervals are used to represent epistemic uncertainty mostly due to missing information, such as IP's fao, ethnicity or subtype of cancer, or facts about the nearest relative with cancer disease.

2.2 Data Reliability Assessment

When it comes to data quality, completeness and access are, of course, the most important issues. This primarily concerns patient data and access to internationally recognized databases Minimum requirements for the comparability of studies are a) a concrete description of the cohort and its characteristics, as well as the objectives of the study; b) the relevant diseases (cancer with type and subtype) from a given classification; c) a concisely described selection of the families and their index patients with regard to disease patterns assigned to one of three risk classes, personal female/male history, current age, age at menopause, age at menarche, child birth history, menopausal status, use of menopausal hormone therapy etc.; d) first and second/third degree relatives to the IP with their medical histories, age, sex and ethnicity;

e) the number of complete FH with identification of affected family members selected for genetic testing; f) a description of the PV testing strategy and type of mutation analysis required depending on FH or population-based screening for founder mutations. Data quality criteria have a direct impact on quality criteria for the other domains, such as accuracy, performance and availability. Let us now explain the notions completeness, consistency, and accuracy by means of examples.

Completeness. In [16, Table 1], the first row is uniquely devoted to FH diseases, and the first column for IPs includes BC_{51+} cases that are not reported in the FH. For example, entry (1, 1) displays no cancer in the IP and no BC_{50-} in the FH with 9 PV carriers of 229 persons, corresponding to rate of 3.9%. The entries on disease patterns IP 1 BC_{51+} & FH 1 BC_{50-} as well as IP 1 BC_{51+} and FH ≥2 BC_{50-} are not consistent in terms of PV risk. Entries for complex disease patterns are missing from Table 3 because no tests were available. Here, a minimum number of cases should be required, which, given the rarity of severe disease patterns in families at low risk, can only be realized in very large cohorts.

In [25] it is reported that "the lowest mutation frequencies were observed in families with three or more cases of postmenopausal BC, but no occurrence of premenopausal BC, OC or male BC," (PV in 20 of 522 families), whereas 109 carriers of 260 in FH with ≥3 OC and no other CA were reported. This is further evidence that in entry (1, 1), which is dedicated exclusively to the occurrence of no cancer in IP and FH, certainly no visible BC_{51+} cases relevant to the RST are included. In other studies, IP family history is missing [29] or incomplete [14].

In Penn's risk model, foa is collected only for BC, but not for groups without BC. Therefore, it is not possible to display PV risk for age-related OC occurrence alone.

Consistency. Data consistency means that the mean values for fao are earlier in carriers of a mutation than in noncarriers. If one arranges disease patterns according to their severity (e.g., earlier onset of disease or additional disease), so are their lBRCAm higher and vice versa. Frank et al. [16] always use median instead of mean values. The median age of first occurrence of a disease in a cohort and the average risk depends on the behavior of the PV risk function at each age level and changes when probands are subjected to a further age restriction, such as limiting cases to individuals under 50 years of age. When only subjects up to 50 years of age are included, the mutation risk for the BC disease cohorts of all ages in Table 1 [16] increases from 15.95% to 18.8%, implying that the median age of first onset falls by approximately 5 years. For a linear PV risk function $f(x){:=}ax{+}b$, with age x, it holds for the mean values $\mathrm{av}(f(x)) = f(\mathrm{av}(x))$. We will assume that, as in the Frank [15] study, foa mean and median of the BC_{50-} group are almost identical.

Buys et al. [9] reported that the majority of women with BC were diagnosed between ages 35 and 59 years. As expected, the positive rate among women who were diagnosed before age 40 was much higher than the average positive rate (9.3%) and ranged from 13% to 18%. Knowing the median age of subjects with positive PV gives an estimate of the mean risk in both directions.

Accuracy. Whereas, in most studies, risk classes for the occurrence of PV are defined relative to explicitly listed disease patterns consisting of individual cases or combinations of BC, OC and their subtypes and to age intervals of first diagnosis, Pujol et al. [36] introduced five adjacent risk intervals $[2\frac{1}{2}\cdot(n-1)\%, (2\frac{1}{2}\cdot n\%]$, n=1,.., 5, for this purpose. On the other hand, the intervals that are obtained by summarizing lBRCAm "to detect BRCA pathogenic of likely pathogenic variant" and the results from a number of studies in the field in Tables 2a and b to characterize the effect of individual BC types in the age groups suggest considerable process uncertainty, which strongly relativizes discrimination among study cohorts across the five risk classes introduced in [36].

Only validated counseling tools, PV risk assessing models and reliable ground truth data from trusted PV risk studies allow the comprehensive computation of an average or interval-based mutation risk for an individual or his/her family members; assignment to average/low, moderate/elevated or high-risk classes and individual counseling; and recommendation for preventive measures by an expert team.

3 Data Management and Quality Assessment in PV Risk Studies

3.1 Databases in the Domain of Pathogenic Variants Related to BC and OC

Approaches for model validation are based on statistical analyses of patient data and their relatives from generally available international databases, preferably with disclosure of the retrieval techniques [36]. Databases enable better understanding of the causes, detection and spread of diseases and optimize treatment and prevention to assist those affected to cure and manage their consequences. Such databases are operated by national and international research institutions, including universities, hospitals, companies and foundations. Experts collect and provide statistical information concerning cancer prevalence in the population including various parameters, such as gender, age, place of residence, profession, disease occurrence, survival expectancy and results of genetic testing. Important examples are the SEER database[5] which incorporates 18 population-based cancer registries or repositories related to national HBOC organizations. Other contents concern the typing and subtyping of cancer, classification of PV found, prevention and standardization of treatment methods. The authors Huang et al. [22] deal with data generation and cloud-based sharing and develop an automatic variant classification and annotation pipeline called CharGer[6] by using internationally adopted guidelines specifically for rare cancer variants. CharGer queries information from ClinVar and GSEA. ClinVar[7] is a freely accessible public archive, the purpose of which is to report "variants found in patient samples, assertions made regarding their clinical significance, information about the submitter, and other supporting data."

GSEA[8] hosts a computational method with underlying database that determines whether an a priori defined set of genes shows statistically significant, concordant differences between two biological states (relationship between mutation sites/location and phenotypes of BC and OC).

GC-HBOC created a panel of experts known as the VUS (variant of uncertain significance) Task Force, which was tasked with reviewing and adapting the classifications of genetic variants in risk genes for hereditary BC and OC based on the most recent data submitted to the central database, as is reported in [38]. Thus, a standardized patient data selection policy can be used in the design of incidence studies of cancers based on germline mutations in BRCA1/2 genes with large groups of participants of various ethnicities.

---

[5] Surveillance, Epidemiology, and End Results https://seer.cancer.gov/

[6] Characterization of Germline Variants https://github.com/ding-lab/CharGer/

[7] https://www.ncbi.nlm.nih.gov/clinvar/variation/55539/

[8] Gene Set Enrichment Analysis https://www.gsea-msigdb.org/gsea/index.jsp

3.2 Data Collection and Analysis in PV Risk Studies

Databases are of particular importance in the validation of risk models for cancer disease and appearance of hereditary pathogenic variants. Metastudies on genetic counseling, genetic testing for BRCA-related cancer and the frequency of germline mutations in families at risk are based on publications in international, peer-reviewed journals and their databases, such as Cochrane Central Register of Controlled Trials and Database of Systematic Reviews, Ovid EMBASE, and MEDLINE Ovid Embase.

Nelson et al. [28] reviewed 103 medical studies and 110 research articles (with 92712 patients in all), including their methodology, scientific rigor, study parameters, relevance, quality criteria and metrics, performance, accuracy and limitations, as well as adverse effects and benefits for the patients.

The authors of [31, 32] examined the validity of models for predicting BRCA1 and BRCA2 mutations and quantified "the accuracy of the following publicly available models to predict mutation carrier status: BRCAPRO, family history assessment tool, Finnish, Myriad, National Cancer Institute, University of Pennsylvania, and Yale University." They assessed the accuracy of these models, which are widely used in clinical and scientific activities, in estimating the probabilities of having a BRCA1/2 PV known for susceptibility to breast and ovarian cancer.

Pujol et al. [36] classified genetic counseling studies into three levels depending on cohort size, (number of probands and families), control and quality assessment of data quality and level of detail of patient information. They derived five levels for likelihood of BRCA1 or BRCA2 mutation (lBRCAm) and three levels of evidence for therapeutic recommendation. Their findings were based on a literature search strategy using variations and Boolean connectors of key terms. They executed a search of the PubMed database for studies published in English between January 1995 and May 2020, using exemplary listed queries of terms related to BRCA clinical testing.

Let us now turn to individual studies that will provide the basis for further investigations in the next chapter.

Frank et al. [16] offered predictions for the occurrence of a spectrum of variants in BRCA1/2-correlated genes classified as deleterious by GenBank [17] with such risk factors as age of onset, personal and FH, and ethnicity (compiled in tables denoted "Frank tables" below). Beginning with a cohort of 10000 individuals from different specified ancestries, several selection processes were performed, finally resulting in two studies with 4716 noAJ and 2233 AJ index patients and their tested family members. The exact reduction process was more complicated: a noAJ group of 7461 individuals is analyzed for BRCA1/2 and an AJ group consisting of 2539 for the three founder mutations, from which were generated two groups with information on FH. As a result, 6724 FH individuals remained in the noAJ group, and 3022 individuals in the AJ group. The median fao for 4663 BC patients was recorded as 44 years of age, and for 779 patients with OC as 53. Based on the results of this survey, the authors identified risk factors for BRCA1/2 mutations and correlated them with rates for developing BC, BC with subsequent OC, and contralateral BC. This allowed them to model the lBRCAm with the help of logistic regression analysis.

Using disease reports from 7352 German families and data from Breast Cancer Linkage Consortium, *Fischer et al.* [14] evaluated the genetic risk models BOADICEA, BRCAPRO and IBIS, as well as the extended Claus model (eCLAUS), which is used to estimate BRCA1/2 PV carrier probabilities and compare their discrimination and calibration. The incremental value of using pathology information in BOADICEA is assessed in a subsample of 4928 pedigrees with available information on molecular receptors.

Kast et al. [25] investigate the prevalence of BRCA1/2 germline mutations in a cohort of 21401 families with suspected pathogenic BRCA variants and BC/OC status of all individual members. Data, including the BRCA1/2 mutation status of the index patient, were collected between 1996 and 2014 in a clinical setting at the time of first counseling using functional analyses and variant classification of the ENIGMA consortium, which contains genetic data from the GC HBOC database.

Based on an analysis of the Japanese HBOC consortium database, *Okano et al.* [29] discussed the prevalence of BRCAx variants and VUS among a Japanese cohort of 2366 individuals who underwent a screening test. They provided the results of 15 international studies published between 2009 and 2020.

Hall et al. [19] examined the frequency of most common BRCA1 and BRCA2 variants including VUS in women of different ethnicities undergoing testing for hereditary breast-ovarian cancer using a clinical database supported by Myriad Genetic Laboratories, Inc.

Buys et al. [9] advocated panel testing to increase the number of women identified as carrying a pathogenic variant in the cohort with various ancestries as opposed to conducting BRCA gene testing alone with a special focus on first age of diagnosis and TNBC status. The frequency of pathogenic variants was stratified by age in five-year increments between 20 and 90 years in all individuals with breast cancer.

What the majority of the studies have in common is that, with cohorts from different ethnicities and different risk classes, they arrive at lBRCAm results within a large range. At the same time, they differ in the completeness of their data on subject selection and on IPs and their FHs. Thus, two questions arise: to what extent are they comparable, and what metrics can be used to reproduce statistical mutation risk scores related to the occurrence of individual or combinations of diseases as accurately as possible? Approaches to achieve this goal are presented in the next section.

3.3 Age-dependent Risk of PV Occurrence in Patients with BC, OC, and OBC

International studies on the frequency of mutations in BC and OC suppressor genes such as BRCA1 and 2 are only comparable if, for the cohorts studied, they report origin (ethnicities) and the period of data collection as well as the databases used, their selection according to the presence of specified disease patterns (average, elevated or high risk), and the number and selection criteria of the individual patients or IPs and their tested family members (FH). The minimum requirement is descriptive statistics, including totals for pre- and postmenopausal first occurrence of BC and OC in common cancers BC, BBC, MBC, TNBC etc., OC, and OBC_{sp} in the same individual or $BCOC_{FH}$ separately for each IP's first- or second-degree relatives together with the first age of diagnosis. Since the risk for mutations of the analyzed genes varies over the considered lifetime—usually assumed to be 20 to 70 years—but increases in certain constellations (TNBC) and ethnicities for patients with younger age, surveys done separately by age groups of five or ten years each lead to more accurate results and better comparability. Unfortunately, the risk curves for the disease patterns are nonlinear and the slopes increase more steeply the earlier the diseases occur. This is particularly true for certain ethnicities such as AJ and for complex disease patterns of BC, MBC, TNBC, BCOC, or OBC_{sp}&OC in FH. If the likelihood lBRCAm is determined only for mean values of fao in the range of 40 to 60 years, linear behavior can be assumed here as a first approximation, and mutual dependencies between the diseases that increase the risk can be neglected. This approach is therefore of limited use in determining accurate mean values for the incidence of pathogenic variants in selected groups if only the frequency of BC and its subtypes, OC, and BCOC and their averaged fao in the age intervals are known.

To find risk bar charts or piecewise linear curves depending on patient's age of first onset we used tables by Frank et al. [16], models for individual and familial mutation risk (Penn II, [33]), or age-depending risk curves separately by cancer type [9] to develop a mathematical expression for estimating the mean BRCA1/2 mutation risk of a cohort across studies and data resources, using piecewise linear mutation risk percent curves for BC, bBC, mBC, TNBC, OC for IP and FH, and the fao in the five linear sections: $BC_{70\text{-}20}$ [4, 4, 7, 11,16, 24] with additional weights 2 and 9 for BBC and MBC, respectively, $TNBC_{70\text{-}20}$[7, 7, 11,16, 24, 26], $OC_{70\text{-}20}$ [3, 3, 7, 11, 15, 19]. $BCOC_{FH}$ is obtained by adding the two weights for BC and OC. For OBC_{sp} in one person, a further weight *sp* [0, 1, 2, 3, 4, 5] is added [5].

For the Ashkenazi ethnicity, in most cases we add a constant to the noAJ risk functions (i.e., 1, 3, 5) and propose MBC_{AJ} 9 + [4,…,9], $BC_{AJ\ 70,20}$ [9, 9, 12, 16, 21, 29] (+5), $OC_{AJ\ 70\text{-}20}$ [8, 8, 12, 16, 20, 24] (+5) and sp_{AJ} [1,2,3,4,5,6] (+1).

Based on these bar charts or piecewise linear curves *ba*(.), *oa*(.) and *sp*(.), which quantify the percent PV risk for the listed cancers at each fao between 20 and 70 years, we calculate the scalar product of the retained frequencies of BC, BBC, MBC, OC, and BCOC in IP and FH members and the related first ages of onset in the intervals of accumulated risk divided by the total number of families or IPs to obtain an average mutation risk lBRCAm for individuals/IPs and their relatives.

With good agreement, the studies' context and their lBRCAm can be replicated and missing data for the FH context of one cohort can even be carried over from another study after scaling the number of diseases BC, OC and BCOC for similar ethnicities, age intervals and risk classes. Below, we report the published lBRCAm scores and our estimates calculated with the same risk curves *ba*(.), *ca*(.), and *sp*(.) found in large international studies on the risk for the occurrence of pathogenic BRCAx variants among index patients and their family members.

Frank study [16]: Results for mean risk values in %: noAJ group: 791/4716 = 16.77, our estimate 16.67 (oe); AJ group: 21.34/20.46 [5]. We compute the lBRCAm without the first entry 9/229 in [16, Table 1] which contains a certain number of not identified FH BC_{51+}. Thus lBRCAm:=782/4487%= 17.43%. Since there is no reported fao mean for BC_{51+} and BC_{40-} diseases, we use for all 6323 BC and 2063 OC the weights $ba(44)=9.4$, $oa(53)=5.8$, and 135 BOC_{sp} together with $sp(44)= 2.6$ and compute lBRCAm = (6323·9.4+2063·5.8+135·17.8)%/4487=73805/4487%=16.45%.

More precisely, working with fao *ba*(55) for BC_{51+} and $ba_{<40}=13.5$, it follows that

$$((2981-862)\cdot ba(44) + 661\cdot ba(55) + 1888\cdot ba(44) + 934\cdot oa(53) + 803\cdot(oa(53) + ba(44)) + 326\cdot oa(53) + 62\cdot(ba(55) + oa(53) + sp(55)) + 73\cdot(ba(44) + oa(53) + sp(44)) + 862\cdot ba_{<40})/4487 = 16.61.$$

Here are the results for the Ashkenazi cohort: 22.35% and our estimation 22.70%, slightly larger because of the first entry with missing $BC_{51+,\ FH}$ not taken into account.

Karst study [25]: lBRCAm: 24.0%, our best estimate: 24.19%.

$$(50688\cdot 7.4 + 6028\cdot 9.4 + 671\cdot 13.6 + 7250\cdot 6.4 + 1917\cdot 15.9)/21401 = 24.193$$

Table 1. Cohort composition in relation to risk factors. For the BC_{40-}, we use *ba*(35) as risk factor. If multiple ethnicities are considered individually, we name the respective group, noAJ and AJ in [16] (from Table 1, 3 therein), African group for [19]. '# families /ip' means the total number of families tested for BRCA1/2 mutations (equal to the number of index patients for some studies) is usually smaller than the total number of persons in the FH. For the subtype TNBC we use $ba_{<45}$=18 [9, Fig.2]. The abbreviation "foa" means average of first age of onset.

Diseases fao $BC_{50-}/_{51+}$/OC	BC_{51+} IP	BC_{51+} FH	BC_{50-} IP	BC_{50-} FH	BC_{40-} TNBC	BBC $_{50-/51+}$ IP	MBC	OC IP	OC FH	OBC_{ip} $_{51+/50-}$	BCOC FH	# families/(ip)
[16] 44/55/53	661	-	1888	2981	862	-	-	326	934	62/73	803	4487
AJ 44/55/53	326	-	578	1140	188	-	-	116	449	19/18	365	1991
[25] 49/-/51½ MBC: 58		24452	-	26236	1267 -	4132 1896	671	-	7250	-	1917 sp=2.1	21401
[29] 47.72/-/50	2054			2081	211		14	89	(143)	62	(268)	2366
[14] 43.3/50.5			6386	9854		1064	88	668	2353	297	-	7352
[19] African ip:40/52½/45 fh:45/57½/55	212	381	1006	1003		-/188 fao: 42.2		25	90	32	271	1767
lBRCAm [ ] our estimation		[16] 16.77/ 17.43 16.61		AJ: 22.35 22.70		[25] 24.02 24.19		[29] 20.2 18.57		[14] 24.1 25.55	[19] 15.6 17.38	

Fischer et al. [14] used data from 7352 IPs and 4927 pedigrees and a lBRCAm of 24.1% to validate BOADICEA, BRCAPRO, and IBIS as well as the extended Claus model (eCLAUS), intended to estimate BRCA1/2 gene mutation probabilities in FH. The question arises: to what extent is the study with probands and nine disease patterns consisting of BC, BBC, MBC, and OC including those with earlier fao qualified to rank the four risk models? The cohort belongs to a category with an elevated gene mutation risk and is similar in composition to the study by Kast et al. [25], in which almost three times as many families were studied and complete family history information was available. Since the absolute frequencies of BC, OC and BCOC disease patterns are not known and slightly varying median values are given instead of means, our calculated lBCRAm = (15240·9.68 + 1064·11.68 + 88·18.68 + 3021·6.8 + 297·19.15) % /7352 = 25.55% is of only limited precision.

Hall [19] presents lBRCAm of PV in BRCAx genes and VUS for 46276 individuals and 7 ethnicities with varying fao:

Western European Group: 36235 families, 12.1%, our estimation 12.91%; African group: 1767 families 15.6%, our estimation: 17.38% using fao 45.2y BC, 48y OC:

212·6¼+1006·11+ 188·12.12+381·4¾+1003·9+25·9+90·5+32·23+271·14=30712

Okano [29]: This database includes 2366 cases in which the probands underwent genetic testing, with 20.1% BRC1/2+ carriers, 2054 female BC patients, 89 OC patients, 62 patients with OBC, and 14 MBC. There are 211 carriers with subtype TNBC and fao=44y in the cohort, from which we extracted 2054 patients with fao 47.7y that had no family history, but had an onset age and subtype information. We used the FH patient data from the Asian cohort in [19] for scaling to produce the missing FH context:

BC: #Index patients/#FH members=766/776; OC: 63/10; BCOC: 31/134;
IP: 1843·7.92+211·18+89·7+62·17.2+14·16.9=14597+3798+623+1066+237=20321
FH: (2081-211)·7.92+211·18+143·7+268·14.92=14810+3798+1001+3999=23608
$lBRCAm_{IP}$=20321/2366% = 8.59%; $lBRCAm_{FH}$=23608/2366% = 9.98%.
[29]: lBRCAm=20.2%, model-based risk estimated 8.59%+9.98%= 18.57%.

The primary uncertainty factor is the lack of data on the first occurrence of the cancer disease and its type and subtype, if applicable. The mean age of first occurrence of a disease in a cohort and the average risk depends on the age distribution of their first diagnoses and the time period considered and changes when probands are subjected to a further age restriction, such as only cases under or over 50 years.

Thus, the mutation risk for the BC disease cohorts of all ages in Table 1 [16] increases from 15.95% to 18.8% when only subjects up to 50 years of age are included, meaning that the median age of first onset falls by approximately 5 years.

3.4 Dempster-Shafer Based Quality Metrics

In [3, 5], we used Dempster's rule to combine data on penetrance of pathogenic gene variants in BRCA1/2 correlated with personal and family history of cancer based on data from [16] for non-Ashkenazi population (cf. Tables 1, 2). Let Ω be the universe with eleven focal elements b_1 proband's (postmenopausal) BC_{51+}, b_2 family member's (premenopausal) BC_{50-}, o_1 (OC_{51+}), o_2 (OC at any age); additional masses for premature diseases: ba_1, oa (fao BC/OC between 40y and 50y), ba_{1+} (BC_{40-}), ba_2 (BC_{40-}), nr (near relative, disease in adjacent generations), bil (bilateral BC), and ma (male BC). Remember that $m(.)$: $2^{\Omega} \rightarrow [0,1]$, $m(\emptyset) = 0$, the mass of empty set (impossible event) is zero and $\Sigma_{X \in 2^{\Omega}} m(X) = 1$, $F(m) = \{X \in 2^{\Omega}, m(X) > 0\}$ set of all focal elements.

A first basic probability assignment BPA with masses m_1 could be based on proband's mutation probabilities (cf. PV risk in Figure 1, first column), a second with m_2 by those of her family members (Figure 1, first row). Using Dumpster's rule with $m_{12}(X) := \Sigma_{X_1 \cap X_2 = X, X_1, X_2 \in 2^{\Omega}} m_1(X_1) m_2(X_2)$, we combine the BPAs for m_1 and m_2 and obtain a combined BPA $m_{DS}(\emptyset) := 0$, $m_{DS}(X) := m_{12}(X) / (1 - m_{12}(\emptyset))$, $X \neq \emptyset$, , whereas the belief function is computed by applying the definition $\mathrm{Bel}(X) := \Sigma_{H \subseteq X} m(H)$.

Table 2. DST-based data fusion for 11 risk factors

Risk Factor %	b_1, b_2	o_1, o_2	$b_1 \cup b_2$, $b_1 \cup o_2$	$b_2 \cup o_1$, $o_2 \cup o_2$	$b_2 \cup o_1 \cup$ o_2, Ω	ba_1, oa_1	ba_{1+}, ba_2	nr, bil, ma	Σ %
m_1	2, 10	7, 7	2, 8	10, 12	2,11	4, 2	5, 6	2, 5, 5	100
m_2	4, 4	4, 4	3, 5	6, 1	1, 39	4, 2	5, 4	4, 5, 5	100
m_{DS}	3.9, 11.9	9.9, 10.2	2.1, 7.1	9.4, 8.6	1.6, 7.7	3.8,1.8	4.8, 5.3	2.3,4.8, 4.8	100
Bel_{DS}			17.9, 21.2	31.3, 28.7	51.7 100				
Ex. 1	x, x	x, x	x, x	x, x	x, -	-, x	-, -	x, -, -	77.4
Ex. 2	-, x	x, -	-, -	x, -	-, -	-, x	-, -	-, -, -	33.1
Ex. 4	x, x	-,x	x, x	-,-	-, -	-, -	-, -	x, -, -	37.5

Example 1: Proband 38y BC and OC, her mother B_{50-} fao≥40y, and her aunt OC $\mathrm{Bel}_{DS} = 77.4\%$ compared to 72.2% [16] and 52% [33].

Example 2: Suppose the patient has OC_{40-} and her aunt BC_{50-}. $\mathrm{Bel}_{DS}(b_2 \cup o_1) = 33.1\%$, 40% [16, Table 2].

This DS approach is based on a study with a high level of evidence that satisfactorily supports the selection of the cohort and its relevant parameters. Due to the rarity of the complex BCOC and OCOC syndromes, the case frequencies are so low that the mutation risks of comparable groups vary considerably. No case numbers are available for BC subtypes with earlier foa. In this respect, the reported beliefs are not reliable lower bounds for the occurrence of pathogenic BRCAx variants and overestimate the risks for randomly selected families and subjects without cancer, but better represent increased risks of common BCOC occurrence than the age-dependent risk curves for (sub)-types of BC or OC.

In summary, the DS schemes constructed on the basis of relevant studies are unsatisfactory at one point or another, either having too few disease categories, having to guess the number of cases in the upwardly unlimited constellations two and more BC_{50-} or OC, or considering only two patients in the FH if the total number of diseases is not given. Moreover, the cohorts include only a small number of families with complex constellations of two or more BC/OC diseases (cf. the enumeration of groups with occurrence of ovarian cancer in [25]) and groups with inconsistent risk likelihoods, particularly striking is the lBRCAm for the proband with BC and a family member with OC and vice versa (which are far apart for the combinations b_1 and o_2 resp. b_2 and o_1; tentatively they should be combined into one constellation).

In order to develop an alternative BPA that is based on the results of highly ranked studies, we define the relevant disease patterns and their risk factors.

These are the sets b_1 and o_1 and b_2 and o_2 for BC and OC occurrence in individuals and family members, ob_1 for one BC and OC in an individual and o_2b_2 for this combination in FH, whereas the sets b_2 and o_2 contain single occurrences but also two and more diseases.

Family History (includes at least one first- or second-degree relative and excludes proband).

Proband	No Breast Cancer < 50 Years of Age or Ovarian Cancer in Anyone		Breast Cancer < 50 Years of Age in One Relative; No Ovarian Cancer in Anyone		Breast Cancer < 50 Years of Age in > One Relative; No Ovarian Cancer in Anyone		Ovarian Cancer at Any Age in One Relative; No Breast Cancer < 50 Years of Age in Anyone		Ovarian Cancer in > One Relative; No Breast Cancer < 50 Years of Age in Anyone		Breast Cancer < 50 Years of Age and Ovarian Cancer at Any Age	
	No.	%	No.	%	No.	%	No.	%	No.	%	No.	%
No breast cancer or ovarian cancer at any age	9/229	3.9	19/434	4.4 b_2	46/419	11	6/153	3.9 o_2	10/117	8.5	58/354	16.4 o_2b_2
Breast cancer ≥ 50 years of age	4/172	2.3 b_1	22/197	11.2 $b_2 \cup b_1$	12/118	10.2	3/69	4.3 $b_1 \cup o_2$	1/18	5.6	19/87	21.8 $o_2b_2 \cup b_1$
Breast cancer < 50/40-49 Year s of age	55/579 16/284	9.5	89/484 31/289	18.4	117/322 41/172	36.3	34/194 15/115	17.5	7/42 3/25	16.7	126/267 55/141	47.2
Ovarian cancer at any age/≥50, no breast cancer	5/77 3/45	6.5 o_1	14/41	34.1 $b_2 \cup o_1$	11/26	42.3	23/83	27.7 $o_2 \cup o_1$	12/28	42.9	38/71	53.5 $o_2b_2 \cup o_1$
Breast cancer ≥ 50 years of age and ovarian cancer at any age	5/27	18.5 ob_1	1/9	11 $b_2 \cup ob_1$	4/11	36.4	1/6	17 $o_2 \cup ob_1$	1/3	33	3/6	50 $o_2b_2 \cup ob_1$
Breast cancer < 50 years of age and ovarian cancer at any age	5/25	20	7/14	50	4/5	80	5/9	56	2/2	100	13/18	72.2

Fig. 1. DS scheme with 15 subgroups built from sets b_1, o_1, b_2 and o_2 for BC and OC occurrence in individuals (first column) and family members (first row), ob_1 for BC and OC in the individual (IP) and o_2b_2 for BCOC in FH, as well as the Cartesian product of these sets.

Finally, we account for the special cases post- and premenopausal BC/OC or premature breast cancer and the subtypes TNBC or bilateral/male BC, with their individual risk factors depending on the fao and the origin of the patient.

These factors are only partially represented in the Frank tables [16] (that is, noAJ or AJ, $BC_{51+/50-/40-}$, $OC_{51+/50-}$), but no subtypes and no BC_{51+} are documented the FH.

Then, we propose the following approach: generalization of Table 1 [16] with 6 × 6 entries, let be $R_{m,n}$ a PV risk matrix with non-negative vector entries at all positions (i, j) for mutually different disease patterns c_{ip}, i=1..m, c_{jf}, j=1..n and individuals p—captions on the left hand side—and their family members/FH f (father, mother, siblings and second-degree relatives on the father's or mother's side)—captions on the upper side. They consist of the number of affected individuals, the number of affected individuals with positive testing, total numbers of associated cancer types to the disease pattern with the age limits within which the first occurrence lies.

A partition of the array into disjoint groups is deemed *consistent* if there is a BPA that assigns all groups in a partition their minimal mutation risk. Here, the belief calculation (to an array entry) must sum the associated masses of the m_1+n_1 groups g_{i_1p}, g_{j_2f}, $1 \leq i_1 \leq m_1 \leq m$, $1 \leq j_2 \leq n_1 \leq n$, built from the first row and column and those of their $m_1 \cdot n_1$ Cartesian set products $g_{i_1p} \times g_{j_2f}$, of disease pattern including BC/OC and possible additional risk factors for earlier fao, or special subtypes of BC (e.g., TNBC etc.).

The masses of the BPAs are defined in such a way that in the belief calculation for a disease pattern in the individual's family, the sum of the associated masses yield only the minimal or, alternatively, the average risk in the corresponding group (subarray of the risk matrix). If the consistency conditions are violated, they can be fulfilled by coarsening the partition while retaining the smaller minima in the new subdivision. That can be useful, for example, for combinations such as $b_f \& o_p$ or $o_f \& b_p$ in the case of Frank's Table 1 [16]. Figure 1 shows a possible partition and Table 3 the corresponding BPAs for minimum and average risk metrics.

Table 3. Minimum and average masses of groups g_{i_1p}, g_{j_2f}, i_1, j_2=1, 2, 3, from the first row and column and those of their nine Cartesian set products $g_{i_1p} \times g_{j_2f}$, that fulfill the equation system for the DS-belief construction.

% min/mean → ↓Differences Δ	o_1=6.5 o_2=3.9/5.9	b_1=2.3/7.85	b_2= /∪noCa 4.4/7.6/6.8	o_2b_2 = 16.4 ob_1 =18.5/19.2	Σ 52/63.2
$b_1 \& b_2$ 10.2/21.4 3.5/5.95/6.75	$o_1 \& o_2$ 27.7/31.5 17.3/19.1	$o_1 \& b_2$ 34.1/37.3 23.2/23.2/24	$o_2 \& b_1$ 4.3/13.9 -/1.15	$o_x \& b_y$ 4.3/15.4 -/1.3,1.65,2.1	44
$ob_1 \& b_2$ 11/41 -/14.2/15	$ob_1 \& o_2$ 17/45 -/19.9	$o_2b_2 \& b_1$ 21.8/41 3.1/16.75	$o_2b_2 \& o_1$ 53.5 30.6/30.6	$ob_1 \& o_2b_2$ 50/66.7 15.1/31.1	48.8
min/av masses	0.065 / 0.059	0.0785	0.076	0.164 / 0.192	
$b_1 \& b_2$ 0.035/0.0595	$o_1 \& o_2$ 0.173/0.191	$o_1 \& b_2$ 0.232/0.224	$o_2 \& b_1$ -/0.015		
$ob_1 \& b_2$ -/0.142	$ob_1 \& o_2$ -/0.199	$o_2b_2 \& b_1$ 0.031/0.168	$o_2b_2 \& o_1$ 0.306/0.306	$ob_1 \& o_2b_2$ 0.151/0.311	Add risk factors *nr*, *ba, ma* and *bil*
$b_{1,51+/50-/40-}$= 0.023/0.095/0.132 $b_{2,50-}$ =0.044/0.076	o_1 =0.065 o_2 =0.059	$ob_{1,51+/50-}$ =0.185/0.2		o_2b_2 0.164	

Example 3: Proband 45y BC and OC, aunt OC

$\text{Bel}_{min}(ob_1 \cup o_2) = m(ob_1) + m(o_2) + m(ob_1 \cup o_2) =$ 0.185+0.039+0 = 0.224, and for the average risk $\text{Bel}_{av}(ob_1 \cup o_2) = 0.192 + 0.059 + 0.199 = 0.45$.

Example 4: Proband BC, sister OC and BC
$\text{Bel}_{min}(b_1 \cup o_2b_2) := m(b_1) + m(o_2b_2) + m(b_1 \cup o_2b_2) = 0.023+0.164+0.031=0.218$
(cf. minimum value of entries (2,6)&(3,6) in [16,Table 1]), and for the average risk
$\text{Bel}_{av}(b_1 \cup o_2b_2) := m(b_1) + m(o_2b_2) + m(b_1 \cup o_2b_2) = 0.0785+0.164+0.1675=0.41$
One can now consider the individual's age of onset and either refine the partition or add a risk surcharge for premenopausal BC.

The results in Table 3 show good agreement with the sizes of prevalence of pathogenic BRCAx variants related to the scores on the cancer diseases checklist used in Ataseven et al., [2, Fig. 2]. The advantage of the proposed methods is twofold: if a BPA can be found with the prevalence study data, it can be used, and study data are consistent and provide reliable assignments of a cohort's family disease patterns to a gene mutation risk class.

Using this DS approach, it is also possible to approximate the mean mutation risk of a cohort separately by BC, OC, or BCOC groups in other studies. We now explore the example of the BC group in Kast [25]. Taking this group discussed in [25, p. 467, Table 3], we select six subgroups with 8026 from 11362 individuals, 1258 pathogenic mutations and a lBRCAm =15.67%, groups that fit into our b_1 and b_2 categories with zero or one BC_{51+} and 1, 2, $\geq$3 BC_{50-}. The related PV risk value results from

$$(1267\cdot 13.7+2577\cdot 8.8+1725\cdot 17.5+1256\cdot 16.2+739\cdot 30.4+462\cdot 27.5)/8026\%= 125740.8/8026\%=15.67\%$$

Assuming our risk factors of 13.2 %, 7.6/6.8%, and 21.4% for the BC_{40-}, b_2, and $b_1 \cup b_2$ subgroups with 1267, 1725, and 5034 families, we find an approximation for $\text{lBRCAm}_{BC}=(1267\cdot 13.2+1725\ 7.6+5034\cdot 21.4)\%/8026=137562/8026\%=17.14/16.97\%$.

As a last example we consider the group $\geq$2BC & $\geq$1 OC with 39.9% prevalence which fits our $o_2b_2\&b_1$ set with lBRCAm = 41%.

4 Conclusions and Further Work

In this work on the validation of risk assessment models for the occurrence of breast and ovarian cancer–related deleterious gene variants, we propose two ways to mathematically model the mutation risk for individuals and their relatives based on disease FH arising from BC and OC. These allow large studies on the frequency of occurrence of pathogenic BRCA variants to be evaluated using data from international databases according to standardized criteria and to calculate the lBRCAm for the whole group or subgroups with different patterns of BC and OC diseases.

The first model is based on risk functions for BC, OC and subtypes depending on the age of first diagnosis, and the second on the calculation of a Dempster-Shafer BPA to given PV frequencies in a partition of the cohort according to occurrence of different BC-OC patterns. In this way, it is possible to predict the lBRCAm for these syndromes or to compare prevalence studies and their results and calibrate them if necessary. The universal metrics reproduce the mean lBRCAm obtained in five large studies with cohorts from different ethnicities and risk classes, suggesting their suitability also for further studies. The second metric, which can be specifically adapted to data from another prevalence study, also allows us to look at comparable subgroups from other studies with their disease patterns and reproduce the mean values for mutation risk. These metrics provide reciprocal statements on their suitability and on quality criteria such as performance, accuracy, and consistency for the risk models, genetic counseling tools and comparative surveys. On the one hand the study results were calibrated correctly, and on the other hand the metrics proposed are universally applicable.

Future work will involve further development of risk functions for the different types and subtypes of hereditary cancers and variety of PVs via access to the relevant databases, consolidation of the data with respect to the quality criteria and automated calculation and visual analytics of their metrics.

Acknowledgement

I would like to express my particular gratitude to my colleague Prof. Ekaterina Auer for her great contributions in the preparation and presentation of the results achieved.

References

1. Amir, E. et al.: Assessing Women at High Risk of Breast Cancer: A Review of Risk Assessment Models. JNCI 102(10) 680-691 (2010) doi: 10.1093/jnci/djq088.
2. Ataseven, B., Tripon, D., Rhiem, K., et al.: Prevalence of BRCA1 and BRCA2 Mutations in Patients with Primary Ovarian Cancer: Does the German Checklist for Detecting the Risk of Hereditary Breast and Ovarian Cancer Adequately Depict the Need for Consultation? Geburtshilfe Frauenheilkd. 80(9) 932-940 (2020) doi:10.1055/a-1222-0042
3. Auer, E. and Luther, W.: Uncertainty Handling in Genetic Risk Assessment and Counseling. J. Univers. Comput. Sci. 27(12), 1347–1370 (2021) doi: 10.3897/jucs.77103
4. Auer, E., Luther, W., Weyers, B.: Reliable Visual Analytics, a Prerequisite for Outcome Assessment of Engineering Systems. Special Issue of the 11th Summer Workshop on Interval Methods. Acta Cybernetica 24(3) 287-314 (2020) doi:10.14232/actacyb.24.3.2020.3
5. Auer, E. and Luther, W.: Dempster-Shafer Theory Based Uncertainty Models for Assessing Hereditary, BRCA1/2-Related Cancer Risk. In: Beer, M., Zio, E., Phoon, K. K., and Ayyub, B. M. (eds.): Proceedings 8th International Symposium on Reliability Engineering and Risk Management (ISRERM 2022), Hannover, Sept. 4-7 (2022)
6. BayesMendel Lab Harvard University—BRCAPRO https://projects.iq.harvard.edu/bayesmendel/brcapro
7. Bellcross, C. A., Peipins, L. A. et al.: Characteristics associated with genetic counseling referral and BRCA1/2 testing among women in a large integrated health system. Genet. Med. 17(1) 43–50 (2015) doi: 10.1038/gim.2014.68. Epub 2014 Jun 19
8. BOADICEA: Centre for Cancer Genetic Epidemiology, Cambridge University https://ccge.medschl.cam.ac.uk/boadicea/
9. Buys, S., Sandbach, J., Gammon, A. et al.: A study of over 35,000 women with breast cancer tested with a 25-gene panel of hereditary cancer genes. Cancer 123(10) 1721–1730 (2017) doi: https://doi.org/ 10.1002/cncr.30498.
10. Cai, L., Zhu, Y.: The Challenges of Data Quality and Data Quality Assessment in the Big Data Era. Data Science Journal 14(2) 1-10 (2015) doi: 10.5334/dsj-2015-002
11. Claus, E. B., Risch, N. et al.: Autosomal dominant inheritance of early onset breast cancer: implications for risk prediction. Cancer 73(3) 643–651 (1994) doi:10.1002/1097-0142(19940201)73:3<643::aid-cncr2820730323>3.0.co;2-5
12. ClinGen. Gene-disease validity. https://www.clinicalgenome.org/curation-activities/gene-disease-validity/
13. Evans, D. G. R., Eccles, Rahman, N. et al.: A new scoring system for the chances of identifying a BRCA1/2 mutation outperforms existing models including BRCAPRO," Journal of Medical Genetics 41(6) 474–480 (2004) doi: 10.1136/jmg.2003.017996.

14. Fischer, C., Kuchenbäcker, K., Engel, C. et al.: Evaluating the performance of the breast cancer genetic risk models BOADICEA, IBIS, BRCAPRO and Claus for predicting BRCA1/2 mutation carrier probabilities: a study based on 7352 families from the GC HBOC. J. Med. Genet. 50(6) 360-367 (2013) doi: 10.1136/jmedgenet-2012-101415.
15. Frank, T. S., Manley, S. A. et al.: Sequence analysis of BRCA1 and BRCA2: Correlation of mutations with family history and ovarian cancer risk. J. Clin. Oncol. 16(7) 2417-2425 (1998) doi: 10.1200/JCO.1998.16.7.2417
16. Frank, T. S. et al.: Clinical Characteristics of Individuals with Germline Mutations in BRCA1 and BRCA2: Analysis of 10,000 Individuals. J. Clin. Oncol. 20(6) 1480-1490 -- (2002) doi: 10.1200/JCO.2002.20.6.1480
17. GenBank genetic sequence database : https://www.ncbi.nlm.nih.gov/genbank/
18. Gupta, S. et al.: Prevalence of BRCA1 and BRCA2 Mutations among Patients with Ovarian, Primary Peritoneal, and Fallopian Tube Cancer in India: A Multicenter Cross-Sectional Study. JCO Glob. Oncol. 7(6) 849-861 (2021) doi: 10.1200/GO.21.00051.
19. Hall, M. J., Reid, J. E., Burbidge, L. A. et al.: BRCA1 and BRCA2 mutations in women of different ethnicities undergoing testing for hereditary breast-ovarian cancer. Cancer 115(10) (2009) 2222–2233 doi: 10. 1002/cncr.24200.
20. Himes, D., O. et al.: Breast cancer risk assessment: Evaluation of screening tools for genetics referral. J Amer. Assoc. of Nurse Practioners 31(10) 562-572 (2019) doi: 10.1097/JXX.0000000000000272
21. Hirst, J., Crow, J., Godwin, A.: Ovarian Cancer Genetics: Subtypes and Risk Factors. https://www.intechopen.com/chapters/58601 doi: 10.5772/intechopen.72705
22. Huang, K. L., Mashl, R. J., Wu, Y., et al.: Pathogenic Germline Variants in 10389 Adult Cancers. Cell. 173(2) 355-370.e14 (2018) doi:10.1016/j.cell.2018.03.039
23. Jervis, S., Song, H., Lee, A. et al.: Ovarian cancer familial relative risks by tumour subtypes and by known ovarian cancer genetic susceptibility variants. Journal of Medical Genetics 51,108-113 (2014) doi: 10.1136/jmedgenet-2013-102015.
24. Karst, K., Schmutzler, R., Rhiem, K. et al.: Validation of the Manchester scoring system for predicting BRCA1/2 mutations in 9,390 families suspected of having hereditary breast and ovarian cancer. Int. J. Cancer: 135(10) 2352-2361 (2014) doi: 10.1002/ijc.28875.
25. Kast, K. Rhiem, K., Wappenschmidt, B. et al.: Prevalence of BRCA1/2 germline mutations in 21401 families with breast and ovarian cancer. J. Med. Genet. 53(7) 465–471 (2016), doi: 10.1136/jmedgenet-2015-103672
26. Kuchenbaecker, K. B., Hopper, J. L., Barnes, D. R.et al.: Risks of breast, ovarian, and contralateral breast cancer for BRCA1 and BRCA2 Mutation Carriers. JAMA 317(23) 2402–2416 (2017) doi: 10.1001/jama. 2017.7112.
27. Mattocks, C. J., Morris, M. A., Matthijs, G. et al.: A standardized framework for the validation and verification of clinical molecular genetic tests: European Journal of Human Genetics 18(12) 1276-1288 (2010) https://doi.org/10.1038/ejhg.2010.101
28. Nelson, H. D., Pappas, M., Cantor, A. et al.: Risk Assessment, Genetic Counseling, and Genetic Testing for BRCA-Related Cancer in Women−Updated Evidence Report and Systematic Review for the US Preventive Services Task Force. Clinical Review & Education 666-685 JAMA. 322(7) 666-685 (2019) doi: 10.1001/jama.2019.8430.
29. Okano, M., Nomizu, T., Tachibana, K. et al.: The relationship between BRCA-associated breast cancer and age factors: An analysis of the Japanese HBOC consortium database. J. Hum. Genet. 66(10) 307–314 (2021) doi: doi.org/10.1038/s10038-020-00849-y.
30. Owens, D. K. et al.: Risk Assessment, Genetic Counseling, and Genetic Testing for BRCA-Related Cancer: US Preventive Services Task Force Recommendation Statement. JAMA, 322(7) 652–665 (2019), doi: 10.1001/jama.2019.10987
31. Panchal, S. M. et al.: Selecting a BRCA risk assessment model for use in a familial cancer clinic. BMC Med. Genet. 9(12) 116-124 (2008) https://doi.org/10.1186/1471-2350-9-116
32. Parmigiani, G. et al.: Validity of Models for Predicting BRCA1 and BRCA2 Mutations. Ann. Intern. Med. 147(7) 441–450 (2007) doi: 10.7326/0003-4819-147-7-200710020-00002
33. Penn II Risk Assessment Model: https://pennmodel2.pmacs.upenn.edu/penn2/

34. Peshkin,B. N. et al.: Genetic testing and management of individuals at risk of hereditary breast and ovarian cancer syndromes. Wolters Kluwer, www.UpToDate.com, 2020.
35. Petrucelli, N., Daly, M., and Pal, T.: BRCA1- and BRCA2-Associated Hereditary Breast and Ovarian Cancer. GeneReviews (1998) updated 2016 Dec 15. [Online]. Available: https://www.ncbi.nlm.nih.gov/books/ NBK1247/.
36. Pujol, P., Barberis, M., Beer, P. et al.: Clinical practice guidelines for BRCA1 and BRCA2 genetic testing, European Journal of Cancer, vol. 146, pp. 30–47, 2021, ISSN: 0959-8049. doi: https://doi.org/10.1016/j.ejca. 2020.12.023
37. Teller, P., Hoskins, K. F., Zwaagstra, A., et al.: Validation of the pedigree assessment tool (PAT) in families with BRCA1 and BRCA2 mutations. Ann. Surg. Oncol. 17(1) 240-246 (2010) doi: 10.1245/s10434-009-0697-9.
38. Wappenschmidt, B. et al.: Criteria of the German Consortium for Hereditary Breast and Ovarian Cancer for the Classification of Germline Sequence Variants in Risk Genes for Hereditary Breast and Ovarian Cancer. Geburtshilfe Frauenheilkd 80(4) 2020, pp. 410- 429; doi: 10.1055/a-1110-0909. https://pubmed.ncbi.nlm.nih.gov/32322110/

Evaluation of Cancer and Stroke Risk Scoring Online Tools

Nelson Baloian[1[0000-0003-1608-6454]], Wolfram Luther[2[0000-0002-1245-7628]], Sergio Peñafiel[1[0000-0002-0025-7805]], Gustavo Zurita[3[0000-0003-0757-1247]]

[1] Department of Computer Science, University of Chile, Santiago, Chile
[2] University of Duisburg-Essen, Germany
[3]School of Economics and Business, Department of Management Control and Information Systems, University of Chile, Santiago, Chile
nbaloian@dcc.uchile.cl wolfram.luther@uni-due.de
penafie@dcc.uchile.cl gzurita@fen.uchile.cl

Abstract. In this work, we present three online tools that calculate the individual and familial risk of stroke, for the occurrence of pathogenic variants in BRCA1 (BReast CAncer 1 gene) or BRCA2 genes with impact on early breast and ovarian cancer disease, and for low and high stage prostate cancer. The forms collect information about the individuals, their specific disease patterns, medical examination results, and the lifestyle of the proband and his/her relatives. Data and model quality and cross-cutting issues such as uncertainty and usability will be addressed in the context of work in which the authors have been involved.

Keywords: Pathogenic Variant, Cancer Risk Web Calculator, Quality Metrics, Risk Model Assessment.

1 Introduction

Cancer and cardiovascular disease account for most premature deaths in the population. While prior diseases and lifestyle are mainly responsible for the latter, the development of cancer has genetic, hormonal and environmental causes. Somatic variants are the most common cause of cancer, occurring from damage to genes in an individual cell during a person's life and are responsible for breast cancer (BC) of various subtypes. While pathogenic variants (PV) interfere with the repair mechanism of altered cells and lead to early disease, as in the case of breast and ovarian cancer (OC), spontaneous mutations occur with increasing age due to hormonal factors or specific risk factors in lifestyle, e.g. in diet or leisure behavior. Germline variants are the cause of 5% to 10% of all cancers. They occur in reproductive cells and are passed directly from the parent to children. Women with a hereditary mutation in the BRCA1 or BRCA2 gene have a six times higher lifetime risk (70%) of developing BC than non-carriers. OC is roughly speaking six times less common than BC for women in the general population. Statements on risk refer to the probability for a person or his/her family to develop a specific form of the disease within a certain period of time or to have a genetic, physical or environmental disposition to more frequent, earlier occurrence or severe progression [1, 4, 5].

Prostate cancer is the most common cancer in men around the world, but age-standardized disease rates vary widely by country and ethnicity, with 34 and 124 cases in

Greece and the United States, respectively, and 185 African-American men per 100 000 in 2016[1].

To assist those affected and their families, medical institutions and governmental organizations have established an extensive education and screening program that allows people to obtain information from primary care physicians, health insurance companies, local pharmacies, or through online offerings, to determine their risk for certain common cancers and, in consultation with individuals they trust, to modify their behavior regarding the type and frequency of screenings, lifestyle, dietary and recreational choices, or to take direct action to reduce their risk [3].

In this post we would like to present three online tools in the form of questionnaires that show the user a percentage statement about their personal or family risk based on an adequate model or algorithm. In doing so, we point to studies and mathematical methods used for risk modeling and to derive the predictions. An essential role is played by databases that store important information on the classification of diseases by type and subtype, stages of disease, appropriate treatments according to the patients, their origin, age, sex, time of first diagnosis, results of screening examinations, comparable disease patterns in the family, that is, first-, second- and third-degree maternal or paternal relatives, and that are maintained and made available by national or international institutions.

Validation of the databases, risk models and calculations to determine the individual or family risk, as well as the presentation of results, interpretation and elaboration of recommendations is the responsibility of the stakeholders, i.e. medical experts, software engineers, users and institutions involved. It is done on the basis of agreed quality criteria and their metrics as well as comparative assessment procedures on the level of evidence and reliability of the studies, tools and resulting recommendations for the patients with suggestions for further prevention, treatment and care [8, 13].

2 Risk Calculators

The PENN II risk model[2] predicts the pretest probability that an individual has a BRCA1 or BRCA2 variant. It was released in a revised version in 2008, is an easy to use internet survey form with ten concrete questions about family (medical) cancer history (i.e. about ancestry, age at first diagnosis, the occurrence of different disease patterns from (bilateral) breast, ovarian, pancreatic and prostate cancer) on one side of the family. The tool allows to compute the likelihood of BRCA1/2 pathogenic mutation (lBRCAm) of the patient/proband or the closest 1st, 2nd relative with cancer (individual or family risk) which results in an increased rate of disease at early age.

The performance for PENN II risk calculator was assessed by the area under the receiver operating characteristic curve (AUC) of sensitivity versus 1-specificity, as a measure of ranking of five prediction tools [6]. In Penn's risk model, only first age of onset for BC is collected, but not for pattern without BC. Therefore, it is not possible to display PV risk for age-related OC occurrence alone. In [3] we used Penn II to build age-depending risk curves separately by cancer type to develop a mathematical expression for estimating the lBRCAm risk of a cohort across studies and data resources and to validate Dempster-Shafer based risk metrics (DSRM). The universal metrics allow

---

[1] DKG 2021 https://www.leitlinienprogramm-onkologie.de/leitlinien/prostatakarzinom/

[2] https://pennmodel2.pmacs.upenn.edu/penn2/

to reproduce the mean lBRCAm obtained in six large studies with cohorts from different ethnicities and risk classes, suggesting their suitability also for further studies.

Another DSRM, which can be specifically adapted to data from a freely chosen reliable prevalence study, also allows us to look at comparable subgroups from other studies with their disease patterns and reproduce the mean values for mutation risk.

The Prostate Cancer Risk Calculator (PCPTRC) was developed based on surveys of a cohort of 5519 men of African American origin in a placebo group of the Prostate Cancer Prevention Trial (PCPT) funded by the National Cancer Institute (USA) with annual determination of the total prostate-specific antigen (PSA) level consisting of bound and so-called free PSA (fPSA), family history (FH), and digital rectal examinations (DRE). If the PSA level exceeded 4.0 ng/ml, an abnormal DRE was detected, and at the end of the study, a biopsy was offered with high evidence for the presence of cancer. A 2012 update of PCPTRC 2.0[3] calculates a risk estimate based on a logistic regression model for presence of low-grade (Gleason grade < 7) and high-grade prostate cancer in a more user-friendly presentation with information about the user population. The involved data collection consisted in an expanded group of 6664 biopsies from 5826 patients and inclusion of other biomarkers of prostate cancer [12]. Unfortunately, the correlation between PSA and DRE and the proposed distinction of biopsy-based classes *no PCa*, *low-grade*, and *high-grade* cancer is poor; if this characteristic is compounded by missing data on the other parameters PRE and FH, the overlap of uncertainty in the assignment to the three risk classes is very large.

The TNM staging system comprises tumor sizes T1, 2, 3, 4, spreading to the lymph nodes N0, 1, and existence of metastases M0; 1abc at different parts of the body. T1: non-palpable, DRE normal, 1a(b): <(≥) 5% of tissue with cancer, 1c: cells found during needle biopsy with elevated PSA, low-grade PCa. T2: palpable, DRE abnormal, 2a(b): (more than) half of one lobe or less (not both), 2c: involves both lobes. T3: tumor outside prostate, T3a: palpable, unilateral capsular penetration, 3b: spread to the seminal vesicles. T4: tumor has spread to tissues next to the prostate.

In parallel, the Gleason grade (GG) describes the aggressiveness of PCa in form of a pair of numbers assigning two grades (x,y) to Gleason score x+y, the first resp. second most common tissue pattern in order of harmfulness. Thus, the pair GG2 (3, 4) represents less aggressive PCa than the pair GG3 (4, 3) and are both assigned to Gleason 7, whereas Gleason scores 8 (GG4) (4, 4) and higher are more dangerous, and Gleason score 6 (GG1) (3, 3) is still below the threshold for clinically significant csPCa, but beyond benign disease.

However, the distinction between low-grade and clinically significant PCa is fluent and shifting downward with the use of new imaging technologies such as nonparametric MRI, higher core (≥12) biopsies, and evaluation of further biomarkers. Important examples are PCA3, a prostate specific, non-protein coding RNA, T2:ERG gene fusions or the prostate health index PHI derived from total PSA value, fPSA and its precursor enzyme subtype [-2]proPSA. There was also a significant correlation between PCa3 and PHI with the Gleason score on biopsy and high-risk pathology [11].

External validation of PCPTRC for 25,733 biopsies from 10 different cohorts and risk predictions were assessed using AUC, discrimination slopes, chi-square tests for quality of fit, and net benefit decision curves [2]. Based on risk curves of (high-grade) prostate cancer as a function PSA level and modifiers FH+/-, DRE +/- and PCA3 for probands aged 65 or 75 years we present a DSRM basic probability assignment (BPA)

---

[3] https://riskcalc.org/PCPTRC/

that models risk weights as a function of four ages (55, 65, 75, 85), PSA value, and modifiers and provides a belief value for the occurrence of a positive biopsy for several grades of PCa which can be adapted to the respective cohort and its ethnic group by applying a simplified calibration rule.

In the case of diseases caused by hereditary gene mutations, a BPA with masses $m_{1,2}$ could be based on proband's or his/her relatives personal risk or experts' estimates reflecting relevant risk factors. Dempster's rule (DR) combines the BPAs for m_1 and m_2, whereas the belief function values as lower bound are obtained by applying the definition $\mathrm{Bel}_m(X) := \Sigma_{H \subseteq X}\, m(H)$. Using DR with $m_{12}(X) = \Sigma_{X_1 \cap X_2 = X, X_1, X_2 \in 2^\Omega}\, m_1(X_1) m_2(X_2)$, a combined BPA $m_{DR}(\emptyset) := 0$, $m_{DR}(X) := m_{12}(X) / (1 - m_{12}(\emptyset))$, $X \neq \emptyset$ is obtained.

Stroke risk calculators: Richards et al [10] mention that at least 110 stroke as a cardiovascular disease (CVD) risk scoring methods exist. They consist of either a point-based scoring questionnaire that required experts or informed probands to manually identify the applicability of various risk factors criteria and to calculate/interpret the sum of points associated with. Online tools allow to directly filling out a web form that transfers the input into a multivariate expression and thus predicts the stroke risk. A problem is missing or incomplete data when it comes to specific questions about the lifestyle of the respondent(s), such as length and duration of walks or cigarette and alcohol consumption.

In [9] mathematical and computer science methods are presented to calculate and assess the risk of stroke occurrence from the dataset of Japanese Electronic Health Records (EHR) from the Tsuyoyama Hospital, Japan, for subjects. The approach requires the application of rules about the assignment of their test results to ranges used by medical experts to interpret a set of medical test results from blood samples or other body-related indices. For this purpose, the classifications obtained with Dempster-Shafer-Evidence Theory are compared with those obtained with other well-known machine learning methods such as Multilayer Perceptron, Support Vector Machines and Naive Bayes. The rules were validated by both medical literature and human experts.

UCL (University of California, Los Angeles) Stroke Risk Calculator[4] collects personal medical and life style data: sex, age, systolic blood pressure, congoing antihypertensive therapy, smoker or nonsmoker, applicable diagnoses such diabetes, history of myocardial infraction, angina pectoris, coronary insufficiency, intermittent claudication, congestive heart failure, atrial fibrillation or other cardiac disorders like left ventricular hypertrophy on ECG which is facilitated by the help of a medical advisor.

A typical model equation is reported in [5]: the 10-year stroke prediction probability (P) was obtained as follows: $(P) = 1 - S(t)\exp(f(\mathbf{x}, \mathbf{M}))$ and $f(\mathbf{x}, \mathbf{M}) := \beta_1 (x_1 - M_1) + \ldots + \beta_k (x_k - M_k)$. Here, $\beta_1, \ldots, \beta_k$ were the regression coefficients of each risk factor, $M_1 \ldots M_k$ the means of the risk factors, and $S(t)$ was the probability of survival from the stroke at the time of 10 years. The half of the data may be used for model construction and the other for external validation. Actual values are given in [5], Table 1, a concrete evaluation of the function is explained in reference 8 therein. To include epistemic uncertainty we propose to use interval computations or fuzzy arithmetic to evaluate (P)

In order to compare different cancer or stroke risk predictors or calculators two aspects should be considered: performance and interpretability. Performance stands for the correctness of the results of the model, typically metrics such as accuracy, F1-score, and the AUC are used. However it is inaccurate to compare methods trained and tested

[4] https://www.uclahealth.org/stroke/stroke-risk-calculator

on different uncalibrated datasets. This is particularly difficult in healthcare since open datasets are scarce. In [9], Table 4 presents a comparison among different stroke risk predictors as a reference.

The next dimension to be considered is interpretability, which is the ability to explain the decision made by the model while making predictions. Calculators based on formulas or simple rules are inherently interpretable since we know exactly how decisions are made, for example it is known which attributes contribute the most to predictions and how they relate. On the other hand, predictors based on artificial neural networks and deep learning are not interpretable because of the large number of internal parameters and non-linear transformations [7]. Interpretability is important for the trustiness and usability of risk calculators even if a method is reported to have a high accuracy. Stroke specialists will likely use a tool that makes sense to them in their clinical practice, and many of them may not use a tool that does not provide an explanation because it may compromise the patient's health and treatments.

3 Conclusion

References to evaluation procedures and metrics that have been carried out enable a team of experts to assess the quality in terms of accuracy, performance, and usability with the participation of health professionals, patients and their families involved in the implementation, supply or use of this service. However, this project is facing a number of challenges. Changes and tightening in the definition of risk classes over the past 20 years, advances in imaging techniques and reporting systems, gene panel testing, and use of related genomic and biomarkers have a major impact on the models and algorithms underlying the risk calculators. Mixed ethnicities in cohorts, lack of data on patients, their disease and family histories, different standards in digital examinations and biopsies performed may lead to highly variable results and incomparability of risk calculators mainly because of epistemic uncertainty, while the authors cited in the references focus mainly on aleatory uncertainty.

As a conclusion, we would like to derive some minimum requirements for the Risk Scoring Online tools: Since the offers are consulted by various user groups, they should have adequate information about their purpose, their operation and the handling of the output results. After an introduction to the objectives and the user groups addressed, comprehensible information must be available for each particular question, which kind of information is expected about the individual's demographics, lifestyle, health status and family disease history. Depending on the disease pattern, examination outcomes, patients' own medical lab samples, test results and biomarkers data must also be at disposal over a longer period of time, which are usually available from the treating medical specialists. In this respect, the questionnaires should be completed in a collaborative manner and the tools support distributed work. Moreover, experts should be given references to relevant literature on the models, datasets and algorithms used. Finally, the outcome should be designed in such a way that users are provided with appropriate counseling and help services or contact addresses depending on their allocation to a risk group and reference to possible effects of various sources of uncertainty.

References

1. Amir, E. et al.: Assessing Women at High Risk of Breast Cancer: A Review of Risk Assessment Models. JNCI 102(10) 680-691 (2010) doi: 10.1093/jnci/djq088.
2. Ankerst, D. P. et al.: Prostate Cancer Prevention Trial risk calculator 2.0 for the prediction of low-vs high-grade prostate cancer. Urology 83(6) 1362-1367 (2014) doi: 10.1016/j.urology.2014.02.035
3. Auer, E. and Luther, W.: Uncertainty handling in genetic risk assessment and counseling. J. Univers. Comput. Sci. 27(12), 1347–1370 (2021) doi: 10.3897/jucs.77103
4. Kuchenbaecker, K. B., Hopper, J. L., Barnes, D. R. et al.: Risks of breast, ovarian, and contralateral breast cancer for BRCA1 and BRCA2 Mutation Carriers. JAMA 317(23) 2402–2416 (2017) doi: 10.1001/jama. 2017.7112.
5. Lee, Jae-woo, Lim, Hyun-sun et al.: The development and implementation of stroke risk prediction model in National Health Insurance Service's personal health record. Comput. Methods Programs Biomed. 153(1) 253-257 (2018) doi: 10.1016/j.cmpb.2017.10.007
6. Lindor, N.M., Johnson, K.J., Harvey, H. et al.: Predicting BRCA1 and BRCA2 gene mutation carriers: comparison of PENN II model to previous study. Familial Cancer 9(4) 495-502 (2010) doi: 10.1007/s10689-010-9348-3
7. Molnar, C., Casalicchio, G., Bischl, B.: Interpretable machine learning–a brief history, state-of-the-art and challenges. In: Joint European Conference on Machine Learning and Knowledge Discovery in Databases, pp. 417-431 Springer, Heidelberg (2020). doi: 10.1007/978-3-030-65965-3_28
8. Owens, D. K.: Risk Assessment, Genetic Counseling, and Genetic Testing for BRCA-Related Cancer: US Preventive Services Task Force Recommendation Statement. JAMA, 322 (7) 652–665 (2019) doi: 10.1001/jama.2019.10987
9. Peñafiel, S., Baloian, N., Sanson, H., Pino, J. A.: Predicting Stroke Risk with an Interpretable Classifier. IEEE Access 9, 1154-1166 (2021) doi: 10.1109/ACCESS.2020.3047195
10. Richards, A., and Cheng, E. M.: Stroke risk calculators in the era of electronic health records linked to administrative databases. Stroke 44(2) 564–569 (2013) doi: 10.1161/ STROKEAHA.111.649798
11. Terracciano, D., Ferro, M.et al: Beyond PSA: The Role of Prostate Health Index (phi). Int. J. Mol. Sci. 21(4) 1184, 14p.(2020) doi: 10.3390/ijms21041184
12. Thompson, I. M. et al.: Assessing prostate cancer risk: results from the Prostate Cancer Prevention Trial. J. Natl. Cancer Inst. 98(8) 529-534 (2006) doi: 10.1093/jnci/djj131
13. US Preventive Services Task Force et al.: Screening for Prostate Cancer: US Preventive Services Task Force Recommendation Statement. JAMA 319(18) 1901-1913 (2018) doi:10.1001/jama.2018.3710

Sales Goals Planning using Evidence Regression

Nelson Baloian[1[0000-0003-1608-6454]], Belisario Panay[1[0000-0002-1440-8192]], Sergio Peñafiel[1[0000-0002-0025-7805]], José A. Pino[1[0000-0002-5835-988X]], Jonathan Frez[1], Cristóbal Fuenzalida[1[0000-0003-0015-1848]]

[1] Department of Computer Science, Universidad de Chile, Santiago, Chile
{nbaloian,bpnay,spenafie,jpino, jfrez,crfuenza}@dcc.uchile.cl

Abstract. Sales planning is a recurrent activity for retail stores in order to provide the necessary resources for a good operation. This is usually based on prediction models which take as input diverse parameters. Modern technology has made possible to easily and economically register three important parameters which are important for sales planning: foot traffic (number of visitors entering a store), conversion rate (which proportion of the visitors make a purchase), and average value of sales ticket. In this paper we present a model for helping retail managers to plan their future sales based on these three simple parameters. The model has been implemented as a sales planning tool which allows them to answer questions like "how much should the foot traffic improve to attain a certain a certain sales goal and how difficult will it be to achieve this goal". The model is based on the Dempster Shaffer plausibility theory which allows and easy interpretation of the results.

Keywords: Sales Planning, Retail Prediction Model, Dempster-Shafer, Plausibility Theory.

1 Introduction

Sales goals planning is a common activity in brick-and-mortar retail stores [1]. It guides store managers concerning a number of decision variables such as staffing, creating new promotions and/or increase advertising during a given planning period. A convenient planning scenario is one in which the manager assumes a target in terms of expected number of sales and then, she obtains from a certain predicting machine the store performance indicator values enabling to reach this goal. There are three typical store performance indicators: foot traffic, conversion rate and average purchase amount per ticket. Therefore, in the mentioned scenario, the output will be a combination of values for these three indicators. For determining sales and operation planning, managers frequently use prediction models for these indicators [2]. The manager evaluates these indicator values and may consider them to be inconvenient or unfeasible. In such a case, she may start again asking the machine with another sales target as input. The scenario thus involves a simulation cycle ending when the manager is satisfied with the obtained indicator values. She may then make the decisions needed to achieve those values. For instance, she may establish a promotion and advertising campaign to accomplish the indicator values. If these values actually occur later on and the predicting machine is correct, then she should satisfy the expected sales goal.

In this paper, we propose the "machine" indicated above. It consists of two systems. The first one is an algorithm intended to forecast the indicators using evidential regression within two months in the future. The regression model is able to find the most important features for the forecasting; thus, it gives the user a clear interpretation of the prediction process. The second system is a method for planning sales goals based on the model predictions. This method uses the model forecast to estimate the expected foot traffic, conversion rate and amount per ticket of the period. The method uses this information and the historical information of the variation of the three indicators to find the optimum daily variations to these indicators that are the most plausible in order to reach the sales goal and its associated risk.

2 The Prediction Algorithm

The prediction algorithm we chose is a supervised learning model called Weighted Evidential Regression (WEVREG) [3] based on the theory of evidence also called the Dempster-Shafer theory [4], which is a generalization of the Bayesian theory. It is more expressive than classical Bayesian models since it allows us to assign "masses" to multiple samples measuring the degree of uncertainty of the process. The method has a precision as good as or even better than other known methods for the problem at hand. Besides, it is interpretable. The experiments made with the method show that it can be used to make good predictions of the three indicators on a daily basis for up to two months.

The algorithm makes a prediction using a sample set (training set) as evidence. When a value is predicted, a set of samples in the training set are assigned a mass (importance) that represents the similarity or effect of each one on the predicted output. The similarity of every sample is computed using a weighted distance function between a new sample (which is being predicted) and its k-nearest neighbors. Because the algorithm is based on the Dempster-Shafer theory it is possible to measure the uncertainty of a prediction. This uncertainty is computed using the distance of the new sample and the values of the outputs observed in the training set. The uncertainty of the responses is represented in each prediction as an upper and lower limit, the bigger the uncertainty, the bigger is the difference between these limits.

As it has been mentioned, the algorithm uses a distance function. This function is tuned during the training phase of the algorithm, where the weight of each feature in the sample set is learned using gradient descent. This means that when a new sample is predicted, the optimal importance of each one of its k-nearest neighbors is computed. Using these weights learned during the training phase, it is possible to perform feature selection tasks.

The weight of each feature in this distance function represents the importance of the feature to predict the desired outcome. This allows us to identify and discard during the training of the algorithm, any feature that does not contribute any information to the outcome of the predictions.

The advantages of using this algorithm over other well-known supervised learning methods, is the interpretability and the uncertainty that the models offer. First, we can assess the importance of each feature in the input vector to learn the effect the selected characteristics have on the desired outcome. And if we want a more detailed explanation for an outcome, the similarity of passed observations (samples) can be inspected. Also, the uncertainty of the algorithm is represented as an upper and lower limit to each predicted outcome, so when a prediction is computed we can assess if a prediction is "good" or "bad" under the algorithm parameters.

3 Computing the Goal Sales Amount

As already mentioned, the input from the user will be her desired future sales, and the output will be the optimal final values of the indicators needed to obtain such goal (number of visitors, conversion rate and average ticket value). Having a precise set of predictions allows her to decide whether to focus on the number of people entering the store, the number of sale tickets they have to make, or the amount of money people spend on each sale.

We define the amount of sales S as the product between the number of visitors E, the conversion rate C and the average value of a ticket T.

$$S = E \times C \times T$$

From the predictions made by the model, we can estimate the expected values of the variables E, C and T, and thus, get the expected value of S.

A sales goal is a value S_g that is expected to be achieved in the future within a certain period, e.g., a month. The value of S_g is generally larger than the original value of S, so it is necessary to increase/adjust the values of E, C and T to achieve the goal S_g. The main objective of the planning tool is to find the optimal values of E, C and T, given a fixed value of S_g provided by the user.

An optimal strategy is to consider the difficulty associated with each variable to be changed over time, because it cannot be assumed that all variables change their values at the same rate. Since we can access their historical values, we assign a number to each variable that represents the mentioned difficulty, whose values are obtained based on the standard deviation of their respective variables. We will call such values the **weights**, and we will name them W_E, W_C and W_T.

Since the relationship between the amount of sales, visitors, conversion rate and average value of a ticket holds, we will consider their initial and final values to work with the equations (the subscripts i and f will be used to denote the initial and final values of the variables):

$$S = E_i \times C_i \times T_i \quad (1)$$

$$S_g = E_f \times C_f \times T_f \quad (2)$$

Moreover, we define the variations V_S, V_E, V_C, V_T as the increase or decrease that they present respective to their initial values. For example, if the number of visitors increases from $E_i = 1000$ to $E_f = 1200$ (growth of 20%), we say that $V_E = 20\% = 0.2$, i.e.:

$$V_S = \frac{S_g}{S} - 1 \qquad V_E = \frac{E_f}{E_i} - 1 \qquad V_C = \frac{C_f}{C_i} - 1 \qquad V_T = \frac{T_f}{T_i} - 1 \quad (3)$$

From (1), (2) and (3) we get:

$$S \cdot (1 + V_S) = E_i \cdot (1 + V_E) \times C_i \cdot (1 + V_C) \times T_i \cdot (1 + V_T)$$

$$\Rightarrow (E_i \times C_i \times T_i) \cdot (1 + V_S) = (E_i \times C_i \times T_i) \cdot (1 + V_E) \cdot (1 + V_C) \cdot (1 + V_T)$$

$$\Rightarrow (1 + V_S) = (1 + V_E)(1 + V_C)(1 + V_T) \quad (4)$$

This reasoning shows that the problem is a matter of working with the variations of the variables, where the objective will be for these to be inversely proportional to their weights (*i.e.*: the higher their difficulty to change is, the less it is required to vary). Thus, we define such proportionality as:

$$V_E \cdot W_E = V_C \cdot W_C = V_T \cdot W_T = x \ \text{(proportion)} \quad (5)$$

Then, without loss of generality, we start to approach the problem focusing on one of the variables, given the symmetry of the product that defines the equation. We use equation (4) and leave the equation as a function of V_E; taking V_S and the weights as constants (since their values are fixed for the problem).

$$\Rightarrow (1 + V_S) = (1 + V_E) \cdot \left(1 + V_E \cdot \frac{W_E}{W_C}\right) \cdot \left(1 + V_E \cdot \frac{W_E}{W_T}\right) \quad (6)$$

When rearranging the equation and amplifying it by $(W_E \cdot W_C \cdot W_T)$, we get the following third-degree polynomial equation:

$$(V_E \cdot W_E)^3 + (W_E + W_C + W_T)(V_E \cdot W_E)^2 +$$
$$\left(\frac{1}{W_E} + \frac{1}{W_C} + \frac{1}{W_T}\right)(W_E \cdot W_C \cdot W_T)(V_E \cdot W_E) +$$
$$(W_E \cdot W_C \cdot W_T) \cdot (-V_S) = 0 \quad (7)$$

To simplify notation, we define the following constants in terms of the weights:

$$c_1 = W_E + W_C + W_T$$

$$c_2 = \frac{1}{W_E} + \frac{1}{W_C} + \frac{1}{W_T}$$

$$c_3 = W_E \cdot W_C \cdot W_T$$

$$c_4 = -\mathrm{V_S}$$

Then, we replace those values in (7) and use the proportion 'x' from (5):

$$x^3 + c_1 \cdot x^2 + c_2 c_3 \cdot x + c_3 c_4 = 0 \tag{8}$$

The previous equation no longer depends specifically on the variable and weight associated with the number of visitors E, but now it only depends on the proportion met by the three variables and the constants c_1, c_2, c_3 and c_4. It can be easily seen that the problem is generalizable for the three variables in the same way, so the problem gets down to solving the cubic equation as a function of x, for which the general formula of Gerolamo Cardano is used.

When seeing what happens with the discriminant Δ of the equation, it is observed that:

$$\Delta = 18c_1c_2c_3^2c_4 - 4c_1^3c_3c_4 + c_1^2c_2^2c_3^2 - 4c_2^3c_3^3 - 27c_3^2c_4^2$$

By applying the constraints of the real problem and the ranges in which all the values lie, we obtain that $\Delta < 0$. Therefore, the equation has a unique positive solution.

To finish, we use the initial values E_i, C_i and T_i (given by the model predictions), the obtained solution of x, the weights, the relationship mentioned in (5) and the definitions of the variations from (6), to conclude the optimal values of E_f, C_f and T_f:

$$E_f = E_i \cdot \left(1 + \frac{x}{W_E}\right) \tag{9}$$

$$C_f = C_i \cdot \left(1 + \frac{x}{W_C}\right) \tag{10}$$

$$T_f = T_i \cdot \left(1 + \frac{x}{W_T}\right) \tag{11}$$

There are other two possible cases for the algorithm before solving the equations mentioned above, depending on which variable values the user fixes as inputs to the process and which ones remain variable and should be calculated. The first case is when one out of the three variables is fixed, and the second one is when two of them are fixed. Since the problem remains symmetrical because of the product shown in (4), fixing one or two variables will is irrelevant for the algorithm being developed.

First case (one fixed variable). When one of the variables is fixed, we end up with its respective variation as a constant, so its respective weight will not play a role in the equations to follow. Without loss of generality, we will work as if the fixed variable is the ticket average value (T).

$$(1 + V_S) = (1 + V_E) \cdot (1 + V_C) \cdot (constant)$$

Then, just as before, we will replace the value of V_C in terms of V_E from (5) as follows, and will name the new constant as C_T:

$$(1 + V_S) = (1 + V_E) \cdot \left(1 + V_E \cdot \frac{W_E}{W_C}\right) \cdot C_T$$

Rearranging the equation:

$$\left(\frac{1}{W_E \cdot W_C}\right) \cdot (V_E \cdot W_E)^2 + \left(\frac{1}{W_E} + \frac{1}{W_C}\right) \cdot (V_E \cdot W_E) + \left(1 - \frac{(1+V_S)}{C_T}\right) = 0 \qquad (12)$$

It is then easily solved as a quadratic equation in terms of the proportion x from (5), for which we define the following constants:

$$a = \left(\frac{1}{W_E \cdot W_C}\right) \qquad b = \left(\frac{1}{W_E} + \frac{1}{W_C}\right) \qquad c = \left(1 - \frac{(1+V_S)}{C_T}\right)$$

Then the remaining equation $[ax^2 + bx + c = 0]$ has the solution:

$$x = \frac{-b \pm \sqrt{b^2 - 4ac}}{2a}$$

Just as in the cubic formula shown in the previous case, we can make sure we have a real solution by checking the value of the discriminant Δ.

Then, from x we can calculate the values of V_E and V_C by using (5):

$$V_E = \frac{x}{W_E} \qquad V_C = \frac{x}{W_C}$$

And finally, the outputs we are looking for:

$$E_f = E_i \cdot \left(1 + \frac{x}{W_E}\right)$$

$$C_f = C_i \cdot \left(1 + \frac{x}{W_C}\right)$$

$$T_f = (given\ by\ user)$$

Second case (two fixed variables). This is the simplest case, since fixing 2 of 3 variables gives us a linear equation. Without loss of generality, this time we work as if both the conversion rate (C) and the average ticket value (T) are fixed by the user. Since their variations V_C and V_T depend on their respective values, they will be constants too, giving the following equation:

$$(1 + V_S) = (1 + V_E) \cdot (constant) \cdot (constant)$$

We name the first constant as C_C and the second one as C_T, and by rearranging, we obtain:

$$(1 + V_S) = (1 + V_E) \cdot C_C \cdot C_T \quad \Rightarrow \quad V_E = \frac{(1 + V_S)}{C_C \cdot C_T} - 1$$

Therefore, the final values are:

$$E_f = E_i \cdot \left(\frac{(1 + V_S)}{C_C \cdot C_T}\right)$$

C_f = (given by user) and T_f = (given by user)

All the above can be done in the same way for the other variables in case the number of visitors is fixed.

4 The Planning Tool App

We developed a web app to visualize the model. Fields can be filled so the user can give her desired inputs for the sales goal and any of the three optional variables. She can also choose the date range in which the model predicts the values. Figures 1 and 2 show an example of the Sales Goal Planning tool:

To understand the example shown in Fig. 1, we can see how the variables are evaluated in the different equations shown before (equations from (1) to (11)). For this example, the weights are not initially shown and will be derived as results from the equations. First, we write our starting values as:

$$E_i = 7{,}329 \qquad C_i = 0.1912 \qquad T_i = 96{,}843.1 \qquad S = 135{,}706{,}700.88$$

We notice that equation (1) is satisfied, since:

$$135{,}706{,}700.88 = 7{,}329 \times 0.1912 \times 96{,}843.1$$

The chosen/desired sales goal value given by the user as "TOTAL" input is:

$$S_g = 200{,}000{,}000$$

So, we proceed to calculate E_f, C_f and T_f by replacing the values into the equations.

(3) $$V_S = \frac{200{,}000{,}000}{135{,}706{,}700.88} - 1 \approx 0.4738$$

(6) $$1.4738 = (1 + V_E) \cdot \left(1 + V_E \cdot \frac{W_E}{W_C}\right) \cdot \left(1 + V_E \cdot \frac{W_E}{W_T}\right)$$

We can see that the Sales Goal Planning Tool shows the value of V_E below the value of E_f in the first box under Results. The same is done for C_f and T_f with their respective V_C and V_T. Therefore, we have:

$$E_f = 8{,}794 \qquad C_f = 0.2177 \qquad T_f = 104{,}492.65$$

$$V_E = 0.1998 \qquad V_C = 0.1384 \qquad V_T = 0.0790$$

$$\Rightarrow 1.4738 = 1.1998 \cdot \left(1 + 0.1998 \cdot \frac{W_E}{W_C}\right) \cdot \left(1 + 0.1998 \cdot \frac{W_E}{W_T}\right), \text{ cf. (6)},$$

and using (5) it follows

$$0.1998 \cdot W_E = 0.1384 \cdot W_C = 0.0790 \cdot W_T = x$$

$$\Rightarrow \frac{W_E}{W_C} = \frac{0.1384}{0.1998} \approx 0.6927 \qquad \frac{W_E}{W_T} = \frac{0.0790}{0.1998} \approx 0.3954$$

$$1.1998 \cdot (1 + 0.1998 \cdot 0.6927) \cdot (1 + 0.1998 \cdot 0.3954) = 1.473757 \ldots$$
$$\approx 1.4738 = (1 + V_S)$$

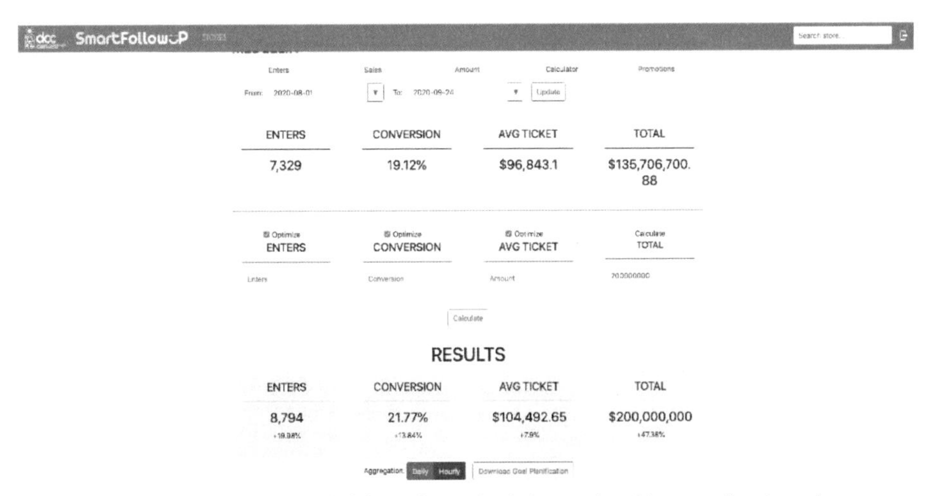

Fig. 1. Predictions for aggregated visitors, input for Sales goal and its associated results.

The purpose of the above equality is to show that the values given as results by the Sales Goal Planning Tool satisfy the equations that are the base for the entire optimization process. i.e.: equations (1), (2), (3), (4) and (5). The same process can be repeated for the other cases where variables are fixed.

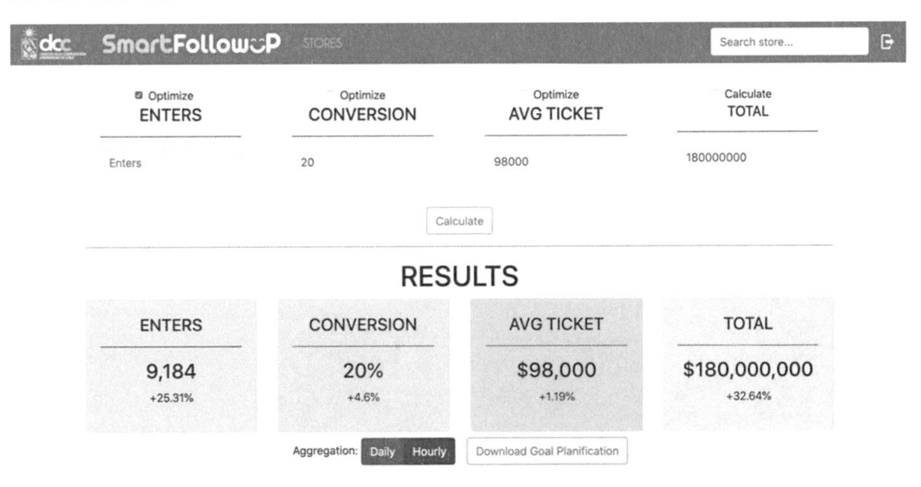

Fig. 2. Same predictions as before, but with 2 fixed variables (C_f and T_f), and a different sales goal with its associated results.

5 Discussion

We concentrated on three store performance indicators. It must be noted that there are many such indicators proposed in the literature. Gopal and Trakkar [5] have broadly classified performance indicators in two groups: financial and non-financial measures. A typical financial performance indicator is Return on Investment [6]. The idea of combining several indicators is well known; an early example of a system combining several of them is the Balanced Scorecard [7]. Among the many performance indicators previously proposed are the following ones, listed by Cai et al. [8]: sales, percent of on-time deliveries, rate of stockouts (losing sales), perfect of order-fulfillment, fill rate, customer satisfaction, order fulfillment lead time, rates of customer complaints, planned process cycle time and cash to cash cycle time.

The three indicators we used (foot traffic, conversion rate and average purchase amount per ticket) were chosen because they were easily obtained from the data we had available. We obtained foot traffic from cameras installed at each store and the other two indicators from foot traffic and sales management software.

References

1. Kreuter, T., Scavarda, L.F., Thomé, A.M., Hellingrath, B., Seeling, M.X.: Empirical and theoretical perspectives in sales and operations planning. Review of Managerial Science 16(3), 319-354 (2021).
2. Pavlyuchenko, K., Panfilov, P.: Application of Predictive Analytics to Sales Planning Business Process of FMCG Company. In, Proceedings of the 13th International Conference on Management of Digital EcoSystems, pp. 167-170. ACM, New York (2021).
3. Panay, B., Baloian, N., Pino, J.A., Peñafiel, S., Frez, J., Fuenzalida, Sanson, H., Zurita, G.: Forecasting key retail indicators using interpretable regression. Sensors 21(5), 1874 (2021).
4. Du, Y.-W., and Zhong, J.-J.: Generalized combination rule for evidential reasoning approach and Dempster–Shafer theory of evidence. Information Sciences 547(2), 1201-1232 (2020).
5. Gopal, P.R. and Thakkar, J.: A review on supply chain performance measures and metrics: 2000-2011. International Journal of Productivity and Performance Management, 61 (5), 518-547 (2012).
6. Anand, N. and Grover, N., Measuring retail supply chain performance. Benchmarking: An International Journal 22 (1), 135-166 (2015).
7. Kaplan, R.S. and Norton, D.P.: The balance scorecard- measures that drive performance. Harvard Business Review 70(1), 71-79 (1992).
8. Cai, J., Liu, X., Xiao, Z. and Liu, J.: Improving supply chain performance management: a systematic approach to analyzing iterative KPI accomplishment, Decision Support Systems 46(1), 512-521 (2009).